The Structural Links between Ecology, Evolution and Ethics

BOSTON STUDIES IN THE PHILOSOPHY OF SCIENCE

VOLUME 296

For further volumes:
http://www.springer.com/series/5710

Donato Bergandi
Editor

The Structural Links between Ecology, Evolution and Ethics

The Virtuous Epistemic Circle

Springer

Editor
Donato Bergandi
Muséum National d'Histoire Naturelle
Centre Alexandre Koyré
Paris, France

ISSN 0068-0346
ISBN 978-94-007-5066-1 ISBN 978-94-007-5067-8 (eBook)
DOI 10.1007/978-94-007-5067-8
Springer Dordrecht Heidelberg New York London

Library of Congress Control Number: 2012954668

Printed on acid-free paper

Springer is part of Springer Science+Business Media (www.springer.com)

dédié à Maouss

Acknowledgements

This book grew out of the international workshop, "Between the Philosophy of Biology and the Philosophy of Ecology: Evolutionisms, Ecologies and Ethics", held in Paris at the Muséum National d'Histoire Naturelle in May 2005. All the contributions except one come from various sessions of the workshop. This event would not have been possible without the financial support and partnership of the French Ministry of Ecology (MEDDE), the Ministry of Foreign Affairs (MAE), the Paris Mayor's Office (Mairie de Paris), the International Union of Biological Sciences (IUBS), and the World Commission on the Ethics of Scientific Knowledge and Technology (UNESCO).

I am thankful to Patrick Blandin for his full collaboration and availability in the realization of this project. I am grateful to Michael Ruse, who, some years ago, accepted that I undertake my post-doctorate, and facilitated my research as a visiting scholar in the Department of Philosophy of Florida State University (Tallahassee). I wish to express my appreciation for the assistance of the Muséum administration, particularly François Bouvier, Brigitte Lebreton, Valérie Teillot and Philippe Cappelle, and for the helpful support of Karen Foulke, Margaret Dancy, Russell Dancy, and Piers Rawling of the Department of Philosophy (FSU). I would also like to thank the FSU library personnel (Dirac, Strozier and Law libraries) for their always present, kind and effective assistance. A very special thanks to Patrick Hamm for his support in helping me to meet the elusive exigencies of English style and grammar.

Contents

Contributors

Robin Attfield is Professor of Philosophy at Cardiff University, where he has taught philosophy since 1968. He has also served as Visiting Lecturer in Philosophy, University of Ife, Nigeria (1972–3), Inter-University Council Visiting Lecturer in Philosophy, University of Nairobi, Kenya, 1975, and National Research Council (Republic of South Africa) Visiting Research Fellow (July/August 1999). He has written the following books: *God and The Secular: A Philosophical Assessment of Secular Reasoning from Bacon to Kant* (1978 and 1993); *The Ethics of Environmental Concern* (1983 and 1991); *A Theory of Value and Obligation* (1987); *Environmental Philosophy: Principles and Prospects* (1994); *Value, Obligation and Meta-Ethics* (1995); *The Ethics of the Global Environment* (1999); *Environmental Ethics: An Overview for the Twenty-First Century* (2003); *Creation, Evolution and Meaning* (2006) and *Ethics: An Overview* (2012). He is the joint editor of *Values, Conflict and the Environment* (1989 and 1996), of *International Justice and the Third World* (1992), and of *Philosophy and the Natural Environment* (1994), and the editor of *The Ethics of the Environment* (2008). For more, see: http://www.cardiff.ac.uk/encap/contactsandpeople/profiles/attfield-robin.html

Donato Bergandi is Maître de conférences in Philosophy of Ecology at the Muséum National d'Histoire Naturelle, Paris. His interests concern the philosophy of biology and the philosophy of ecology (methodology in particular), environmental ethics, environmental philosophy, and the social, economic, environmental and ethical aspects of sustainable development. He has published papers in *Ludus Vitalis*, *Kybernetes*, *Acta Biotheoretica*, *Revue d'histoire des sciences*, and in the collective works *The European Origins of Scientific Ecology* (1998), and *Ecology Revisited: Reflecting on Concepts, Advancing Science* (2011). He participated in setting up the code of ethics in the IUCN *Biosphere Ethics Initiative*. For more, see: http://www.koyre.cnrs.fr/spip.php?article49

Patrick Blandin is Emeritus Professor at the Muséum National d'Histoire Naturelle, Paris. He was Director of the General Ecology Laboratory (1988–1998), Director of the Grande Galerie de l'Evolution (1994–2002), and Director of the Entomology Laboratory (2000–2002). His thesis dealt with spider ecology in an African savannah, under the auspices of the French International Biological Program. Later, he directed interdisciplinary programs on suburban forests and open field woodlots near Paris. He is an internationally known specialist in the systematics and biogeography of *Morpho* butterflies from tropical America, and is currently working on biodiversity issues in northern Peru. He is a Board member of the UNESCO Chair: Développement et Aménagement Intégré des Territoires. He has received awards in France from the Société Zoologique, the Société Entomologique and the Academy of Sciences. He created and chaired the French National IUCN Committee (1992–1999). Since 2003, he has been a member of the Muséum's "Hommes-Natures-Sociétés" Department, where he studies historical, epistemological and ethical issues concerning the conservation of nature and biodiversity. Since 2008, he has been Co-chair of the IUCN *Biosphere Ethics Initiative*. For more, see: http://www.patrickblandin.com/

J. Baird Callicott is University Distinguished Research Professor and formerly Regents Professor of Philosophy at the University of North Texas. He is the co-Editor-in-Chief of the *Encyclopedia of Environmental Ethics and Philosophy* and author or editor of a score of books and author of dozens of journal articles, encyclopedia articles, and book chapters in environmental philosophy and ethics. Callicott has served the International Society for Environmental Ethics as President and Yale University as Bioethicist-in-Residence, and he has served the UNT Department of Philosophy and Religion Studies as chair. His research goes forward simultaneously on four main fronts: theoretical environmental ethics, comparative environmental ethics and philosophy, the philosophy of ecology and conservation policy, and biocomplexity in the environment, coupled natural and human systems (sponsored by the National Science Foundation). Callicott is perhaps best known as the leading contemporary exponent of Aldo Leopold's land ethic and is currently exploring an Aldo Leopold Earth ethic in response to global climate change. He taught the world's first course in environmental ethics in 1971 at the University of Wisconsin-Stevens Point. His teaching at UNT includes graduate and undergraduate courses in ancient Greek philosophy and ethical theory. His monographs include *In Defense of the Land Ethic* (1989); *Earth's Insights* (1994); *Beyond the Land Ethic* (1999); *Éthique de la Terre* (2010); and *Pensées de la Terre* (2011). For more, see: http://philosophy.unt.edu/people/faculty/j-baird-callicott

Jean-Marc Drouin is Professor of History and Philosophy of Science at the Muséum National d'Histoire Naturelle, Paris. His main topics are the naturalist tradition in the life sciences, the relation between natural history and the public, and the connection between philosophy and natural history. His publications include: *Réinventer la nature* (1991; reissued under the title *L'écologie et son histoire*, 1993); *L'herbier des philosophes* (2008); "Ants and Bees between the French Revolution and the Darwinian Revolution", *Ludus Vitalis* (2005). For more, see: http://www.koyre.cnrs.fr/spip.php?article139

Jean Gayon is Professor at University of Paris 1 (Panthéon Sorbonne), Institute of History and Philosophy of Science and Technology. His main interests are the philosophy of science and the philosophy and history of biology (evolution, genetics, molecular biology). He is the author of *Darwinism's Struggle for Survival: Heredity and the Hypothesis of Natural Selection*, Cambridge University Press, 1998. For more, see: http://edph.univ-paris1.fr/gayon.html.

Catherine Larrère is Professor of Philosophy at the University of Paris 1 (Panthéon-Sorbonne). Trained in political and moral philosophy, she is currently interested in ethical and political issues related to the environmental crisis and the development of new technologies: nature conservation, risk assessment, and the ethics of biotechnologies and nanotechnologies. She has published *L'invention de l'économie. Du droit naturel à la physiocratie* (1992); *Actualité de Montesquieu* (1999); *Les philosophies de l'environnement* (1997); *Du bon usage de la nature, Pour une philosophie de l'environnement* (with Raphaël Larrère, 1997 and 2009), and has co-edited *La crise environnementale* (with Raphaël Larrère, 1997) and *Nature vive* (2000). She is a member of the French National Council of Natural Parks and the Joint National Ethics Committee on Agronomical Research. For more, see: http://www.univ-paris1.fr/recherche/page-perso/page/tx_oxcspage personnel_pi1[uid]=larrere&cHash=2ee0dcb4b40deb801ffe9b381c8fcd0f

Bryan G. Norton is Distinguished Professor of Philosophy, Science and Technology in the School of Public Policy, Georgia Institute of Technology. His current research concentrates on sustainability theory and on problems of scale in the formulation of environmental problems. He is the author of *Why Preserve Natural Variety*? (1987); *Toward Unity Among Environmentalists* (1991); *Searching for Sustainability* (2003), and *Sustainability: A Philosophy of Adaptive Ecosystem Management* (2005). Norton has contributed to journals in several fields and has served on the Environmental Economics Advisory Committee of the US EPA Science Advisory Board, and for two terms as a member of the Governing Board of the Society for Conservation Biology. He was a member of the Board of Directors of Defenders of Wildlife from 1994–2005 and is currently on their Scientific Advisory Board. He is also a member of the Advisory Committee for the MacArthur Foundation Advancing Conservation in a Social Context Project. For more, see http://www.spp.gatech.edu/aboutus/faculty/BryanNorton

Tom Regan is Emeritus Professor of Philosophy, North Carolina State University. During his more than 30 years on the faculty, he has received numerous awards for excellence in undergraduate and graduate teaching; was named University Alumni Distinguished Professor; published hundreds of professional papers; wrote or edited more than 20 books; won major international awards for film writing and direction; and presented hundreds of lectures throughout the United States and abroad. In 2000, he received the William Quarles Holliday Medal, the highest honor NC State University can bestow on one of its faculty. For more, see: http://www.lib.ncsu.edu/animalrights/Animal_Rights/Ethics/index.html

Michael Ruse is Lucyle T. Werkmeister Professor of Philosophy at Florida State University, Tallahassee. He began his teaching career at the University of Guelph, where he taught for 30 years. He writes on the nature of sciences, evolutionary biology, evolutionary ethics and the nature of value. Among his most recent book there are: *Mystery of Mysteries: Is Evolution a Social Construction?* (1999); *The Evolution Wars: A Guide to the Debates* (2000); *Can a Darwinian Be a Christian? The Relationship between Science and Religion* (2001); *Darwin and Design: Does Evolution Have a Purpose?* (2003); *The Evolution-Creation Struggle* (2005); *Darwinism and Its Discontents* (2006); *Charles Darwin* (2008); *The Oxford Handbook of Philosophy of Biology* (ed.) (2008); *Defining Darwin: Essays on the History and Philosophy of Evolutionary Biology* (2009); *Science and Spirituality: Making Room for Faith in the Age of Science* (2010); *The Philosophy of Human Evolution* (2012) and as editor *The Cambridge Encyclopedia of Darwin and Evolutionary Thought* (2013). For more, see: http://www.fsu.edu/~philo/new%20site/staff/ruse.htm

List of Figures

Chapter 1
Ecology, Evolution, Ethics: In Search of a Meta-paradigm – An Introduction

Donato Bergandi

Abstract Evolutionary, ecological and ethical studies are, at the same time, specific scientific disciplines and, from an historical point of view, structurally linked domains of research.

In a context of environmental crisis, the need is increasingly emerging for a connecting epistemological framework able to express a common or convergent tendency of thought and practice aimed at building, among other things, an environmental policy management respectful of the planet's biodiversity and its evolutionary potential.

Evolutionary biology, ecology and ethics: at first glance, three different objects of research, three different worldviews and three different scientific communities. In reality, there are both structural and historical links between these disciplines. First, some topics are obviously common across the board. Second, the emerging need for environmental policy management has gradually but radically changed the relationship between these disciplines. Over the last decades in particular, there has emerged a need for an interconnecting meta-paradigm that integrates more strictly evolutionary studies, biodiversity studies and the ethical frameworks that are most appropriate for allowing a lasting co-evolution between natural and social systems. Today such a need is more than a mere luxury, it is an epistemological and practical necessity.

In short, the authors of this volume address some of the foundational themes that interconnect evolutionary studies, ecology and ethics. Here they have chosen to analyze a topic using one of these specific disciplines as a kind of epistemological platform with specific links to topics from one or both of the remaining disciplines. Michael Ruse's chapter, for instance, elucidates some of the structural

D. Bergandi (✉)
Department Men-Natures-Societies, Centre Alexandre Koyré, Muséum National d'Histoire Naturelle, 75005, Paris, France
e-mail: bergandi@mnhn.fr

D. Bergandi (ed.), *The Structural Links between Ecology, Evolution and Ethics: The Virtuous Epistemic Circle*, Boston Studies in the Philosophy of Science 296,
DOI 10.1007/978-94-007-5067-8_1, © Springer Science+Business Media Dordrecht 2013

links between Darwinism and ethics. Ruse analyzes the Evolutionism vs. Creationism debate, emphasizing the risks run by scientists when they ideologize the scientific content of their studies. In the case of the contributions of Jean Gayon and Jean-Marc Drouin, which respectively deal with the disciplines of evolutionary biology and ecology, some central connections have been developed between these two disciplines, while reserving the option to consider in detail their topic in order to discover essential features or meanings. Gayon analyzes the multilayered meanings of "chance" in evolutionary studies and the methodological implications that accompany such disparate meanings. From a similar analytical perspective, Drouin's contribution focuses on the identification and critical evaluation of the different conceptions of time in ecology. Chance and time, factors of evolution in species and ecological systems, play a very important function in both disciplines, and these chapters help to capture their polysemous structure and development. Bryan Norton's chapter, on adaptive environmental management, is set within an epistemological context where the Darwinian paradigm, ecological knowledge and ethical frameworks meet to give rise to practical, conservationist policies. In his contribution, Patrick Blandin pleads for the necessity of an eco-evolutionary ethics capable of fully encompassing humanity's responsibility in the future determination of the biosphere's evolutionary paths. Our value systems must recognize the predominant place that humanity has taken in the evolutionary history of the planet, and integrate the ethical ramifications of scientific advances in evolutionary and ecological studies. The chapter by J. Baird Callicott introduces us to a metaphorical ecological reversion with direct consequences for our moral conduct. If ecology showed that ecosystems are not organisms, recognizing organisms as a kind of ecosystem could be the basis for a new post-modern ecological ethics that lays the foundation for a better moral integration of humans with the environment. The contributions of Robin Attfield and Tom Regan delve into some of the classical issues in environmental ethics, situating them within a broader ecological and evolutionary context. Attfield's chapter tackles the confrontation between individualistic and ecologically holistic perspectives, their different approaches to the issue of intrinsic value, and their tangled relation to monism and pluralism. Regan's contribution ponders the criteria that allow individual beings, human and non-human, to own moral rights, the role of the struggle for existence in the relationship between species, and the logical difficulties involved in attributing intrinsic value to collective entities (species, ecosystems). Catherine Larrère's chapter discusses the opposition between two environmental and ethical worldviews with very different philosophical centers of gravity: nature and technology. These opposing perspectives have direct consequences not only for the perception of the problems at hand and for what entities are deemed morally significant, but also for the proposed solutions.

To set out some foundational events in the history of evolutionary biology, ecology and environmental ethics is a first necessary step towards a clarification of their major epistemological orientations. On the basis of this inevitably non-exhaustive history, it will be possible to better position the work of the different contributors, and to build a meta-paradigm, i.e. a connecting epistemological framework resulting from one common or convergent tendency of thought and practice shared by different disciplines.

1.1 Some Landmarks of an Interweaved History of Ecology, Evolution and Ethics

From the beginning, with Ernst Haeckel (1834–1919), ecology showed an integrative tendency. Unlike biology, which is fundamentally interested in the structures and functions of organisms, ecology examines the conditions of existence of organisms as they are integrated into their environments. As the discipline developed, controversy arose over the fundamental units of nature that should be the focus of ecological research. In plant ecology, for example, an epistemological rivalry began between organicistic and individualistic perspectives. Frederic Edward Clements (1874–1945) considered the plant formation and successively the biome – the plant-animal formation or community – as a unit composed of individuals that are closely interdependent (1905, 199; 1916, 106, 124–125; 1935, 342–343; Clements and Shelford 1939, 20–24; see also Phillips 1931, 19). On the other side, Henry Allan Gleason (1882–1975), postulated a continuing variation in space and time of the plant association. He maintained that a fixed and definite vegetational structure does not exist and that the plant association was a fortuitous juxtaposition of individuals, a coincidence resulting from environmental selection of the available immigrant species (Gleason 1917, 464, 467, 480; 1926, 15–16, 23–26). In the midst of this epistemological battle for hegemony in the discipline, Arthur George Tansley (1871–1955) introduced the concept of the ecosystem. From the point of view of the ecologist, the ecosystem was conceived as "the basic unit of nature," and, as this methodological abstraction was more integrative and systemic than the entities that had preceded it, Tansley's concept enabled modern ecology to flourish (1935, 299; see: Bergandi 2011).

In animal ecology, with Alfred James Lotka (1880–1949), Vito Volterra (1860–1940), Charles Sutherland Elton (1900–1991), Georgii Frantsevich Gause (1910–1986), Umberto D'Ancona (1896–1964), and in limnology with Raymond Laurel Lindeman (1915–1942), ecology ceased to be an exploratory and descriptive discipline and became a modern experimental science focused on the various functions of the ecological systems at different levels of complexity. Elton (1927) identified some basic principles of organization of animal communities, stressing the importance of the size and numbers of the animals, and of the flow of matter and energy through different levels of consumption (food-chains, food cycles, pyramid of numbers, niche: for the concept of the niche, see Grinnell 1917; Whittaker and Levin 1975). Lotka (1925), Volterra (1926), Gause (1934), Volterra and D'Ancona (1935), D'Ancona (1939) developed mathematical models to study the struggle for existence among animals and to establish laws describing the multiplication of organisms. These models proposed a mechanistic view of the effects of individual organisms on the population aggregate (see Chap. 4 by Drouin, in this volume). Upon the basis of Elton's principles and Tansley's concept of the ecosystem, Lindeman, focusing on food-cycle relationships and his concept of "dynamic ecology," proposed a generalization of thermodynamics based on the exchange of energy between living beings at various trophic levels of an ecosystem (1942, 399–400, 409, 415; see also 1941). Subsequently, Herbert George Andrewartha

(1907–1992) and Louis Charles Birch (1918–2009) proposed a functional concept of the environment centered around the ecological web, where an animal's environment consists of everything, living and non-living, that might directly or indirectly influence its chance to survive and reproduce (1954, 17–24; 1984, 3–18).

These are some of the epistemological cornerstones of three major orientations in ecology: population and community ecology, evolutionary ecology and ecosystem ecology (for global and landscape ecology see Bergandi 2000). As far back as the 1940s, ecologists were looking for a new paradigm that unified evolution and ecology. Warder Clyde Allee (1885–1955), Orlando Park (1901–1969), Alfred Edwards Emerson (1896–1976), Thomas Park (1908–1992), and Karl Patterson Schmidt (1890–1957), traditionally identified as the *Chicago School* of ecology, dealt with the problem of aggregation and animal cooperation on different phylogenetic levels. Their work focused on the connection between ecology and the theory of Darwinian selection, and the link between ecology and genetics was clearly represented as the foundational element of future research in evolutionary ecology. Their proposition, grounded on a population-system approach and on a predilection for group selection considered as the fundamental engine of evolution, anticipated what has come to be known as the units of selection issue (1949, 5, 6, 8, V sect.). Some decades later, this would take the form of a harsh scientific and epistemological confrontation between proponents of group (Wynne-Edwards 1962) and individual selection (Williams 1966).

George Evelyn Hutchinson (1903–1991) made another remarkable attempt to durably connect evolution and ecology when he asked: "Why are there so many kinds of animals?" Anticipating the present issue of biodiversity, his answer connected ecological and evolutionary considerations and pointed out that the genetic variability of a small population of a species will necessarily be lower than in a larger one. He concluded that a diversified community would be better able than an undiversified group to seize new evolutionary opportunities (1959, 155). However, only relatively recently have ecology and evolution merged into a productive scientific field, evolutionary ecology – an ecology that makes explicit what was implicit in *On the Origin of Species*, i.e., the acknowledgement that natural selection is deeply rooted in ecological processes. George Christopher Williams's epistemological campaign against group selection, interpreted as a more onerous hypothesis than individual selection, for a long time largely contributed to the shaping of evolutionary biology. Instead, evolutionary ecology, from the beginning, had a more nuanced position in regard to group selection. In the wake of Allee and his colleagues, Vero Copner Wynne-Edwards (1906–1997), and above all, Lewontin (1970), with his critique of selection units, and in consonance with the interdemic group selection model (Wilson 1975, 1983; see also later: Sober and Wilson 1998) Pianka, considered that group selection actually occurs, even if less frequently than individual selection (Pianka 1974, 13; see also: Emlen 1973, 38–42; Wilson 2001; for the marginality of group selection and the preponderance of kin selection see Hamilton 1964; Maynard-Smith 1964, 1976, 1998).

Phenomenologically speaking, modern population and community ecology built themselves on overcoming the organismic Clementsian idea of community. They harnessed an individualistic Gleasonian perspective on the plant association, and showed that species associations vary, constantly and continuously, in space and in time (Whittaker 1956; Strong et al. 1984; Roughgarden and Diamond 1986). However, the Clements-Gleason epistemological confrontation also had a deeper significance, as both perspectives anticipated structural aspects of certain tendencies in the development of ecology. One current of development essentially focused on the general patterns and functions of ecological systems, while another was structured around analytical, merological-mechanistic models that define the analyzed system with equations so as to make predictions about its behavior, or at least, to explain its structure and dynamics. The latter mechanism-oriented current considers the attributes of communities as resulting from the study of their component populations as well as individuals and their interactions. In other words, the community system is decomposed into lower-level components as populations and individuals in a search for concepts and considerations that belong to the behavioral, physiological or morphological levels (Hutchinson 1965, 110; Price 1986, 3–4; Schoener 1986, 99–100; Inchausti 1994, 213 ff.; see also Kingsland 1985, 50 ff.). The former macro-level pattern-oriented current, Odum's systems ecology, is predominantly interested in focusing on the functional characteristics of ecological systems, using cybernetics models that describe energy flows and nutrient cycles, i.e., energy and materials transfer between the various trophic levels of the ecosystem (Simberloff 1982, 87; McIntosh 1980, 204–205; 1985, 203–208; 1987, 321, 334–336; Hagen 1992, 136–138; Bergandi 1995, 154–168).

During the same period in which the population view of evolution represented the epistemological reference point in evolutionary ecology (Allee et al. 1949), Ronald Aylmer Fisher (1890–1962) worked on constructing a theoretical and methodological link between Darwinism and Mendelism. Mendelian inheritance, which involves the segregation of factors and not their blending, holds that individual genes pass from generation to generation entire and unchanged (1924, 202), and that they constantly tend to create genetic situations favourable to their survival (1930, 95). In a similar vein, Williams later argued that natural selection ultimately arises from reproductive competition among the genes (1966, 251). This was the epistemological groundwork upon which Richard Dawkins proposed his gene's-eye view of evolutionary processes, with its corollary of the "selfish gene" theory. His "genes-replicators" are competing directly with their alleles for survival, are able to create copies of themselves, and program organisms as survival machines to safeguard their existence (1989, 2, 15, 19, 35, 36, 98 [1976]). David Hull recognized the validity of Dawkins's concept of the replicator, but considered that natural selection results from the necessary interplay of the processes of replication and interaction. He thus proposed the concept of the "interactor" to accompany the replicator. The interactor was meant to be a cohesive entity interacting with its environment so that replication is differential (1980, 318, 1998, 150; 2001, 22, 38 [1988]).

The gene-centered worldview of evolution has been firmly contested by many scholars, among them Susan Oyama. Her developmental system theory refuses the informational gene concept, according to which the genes intrinsically contain programs, plans with a predetermined formative power that generate specific organism traits. More particularly, Oyama points out that the genetic imperialism dominant in developmental and evolutionary studies involves an asymmetric dichotomy between the causal values of genes and the environment. The organism-environment complex renders impossible any attempt to argue that in phylogeny and ontogeny the genes represent the primary causal factors (2000, 67–68, 107, 197–198 [1985]; for critiques of the Dawkins gene-centered evolutionary worldview see: Hull 2001, 32 [1988]; Lewontin 1991, 48; Griffiths and Gray 1997, 487; Godfrey-Smith 2000, 411–412; Morange 2001, 159–163, 167; Okasha 2006, 166–172).

In moral philosophy, the meeting between evolutionary and ecological thinking set off a major revolution in ethics: the recognition of an ontological continuity between humans and nature and the concern for an ethics capable of integrating this new perception of humanity's place in the world. A new ethical domain has arisen: environmental ethics. What are the limits of moral community? What entities are worthy of being recognized as bearing intrinsic value? Or to put it differently, do we have ethical duties only to humans or do we also have direct duties to environmental entities? The intrinsic value issue plays a paramount role in channeling analysis about the different types of human relationships to the rest of nature. In a very propaedeutic way, there are at least three accepted meanings of the syntagma "intrinsic value," i.e. non-relational, non-instrumental, and objectivist meanings. We can judge an entity to be endowed with intrinsic value when its value is dependent solely on some inherent properties belonging to the entity in question, or when its value depends solely on the intrinsic nature of the entity (Moore 1922, 260). We can consider an entity as having intrinsic value when it is an end in itself, or when we recognize it solely for its own sake and not as a means to an end (Kant 1990, 45–46 [1785]; see below on Kant's ethics). Finally, we can say that an entity has intrinsic value when its value is not dependent on the observer's perception, appreciation or evaluation, or when its properties or qualities belong to it independently of its being valued (Dewey 1944, 452; O'Neill 1992, 120).

Presently, in Western cultures, according to our ethical and juridical norms, the environment is not judged to have intrinsic value or directly to possess moral, or legal, rights. The environment can be protected only indirectly through the exercise of the rights of human beings or with specific legislation that freezes, in a conservationist sense, the traditional relation that people have with the environment. This traditional relation is primarily economic, and expresses itself in the exercise of property rights. Nowadays, this legal perspective is grounded in a secular hierarchical ethical framework. Hierarchicalism, the origins of which are religious, maintains that things and norms are ordered along a scale of good, with higher and lower norms. One of the central principles of the Great Chain of Being framework is the following: "persons are more valuable than things" (Geisler 1971, 115). Self-awareness, self-determination, and inter-subjectivity (the power to relate to others) are specific characteristics of the human species, and these consciousness- and reason-oriented specificities are at the foundation of Western ethics.

Human-centered ethical perspectives consider that humans alone deserve moral respect. People are the only beings that can have interests, actually or potentially, and thus moral rights (McCloskey 1965, 126–127). "... What is good for humans is, in many respects, good for penguins and pine trees." Humanity must strive to find a balanced cost-benefit approach to the problems of pollution and resource exploitation (Baxter 1974, 5–6, 8–9). The ecological links between people and other components of the ecosystem certainly form a community, but not a "moral community" established on mutual obligations, common interests, and shared rights: "the idea of 'rights' is simply not applicable to what is non-human" (Passmore 1974, 116). A softer anthropocentric option maintains that wild species and ecological systems, even if not endowed with intrinsic value, have "transformative value." Their existence allows the emergence in humanity of a higher level of consciousness, of a perceptual and conceptual shift in ethical worldview that involves less consumptive and destructive habits (Norton 1987, 10–14, 233–239; 2003, Chap. 2, particularly, 32–33; 2005, Chap. 6). Others hold that to cope with the environmental crisis no new ethics is needed, and that it is sufficient to follow an ecologically-updated version of secular, or religious, stewardship traditions. Humans are stewards of nature, where any entity that has a good of its own has moral standing, but not necessarily intrinsic value, and equal moral significance in case of a conflict of interests. This is an ethics marked by caution towards anthropogenic transformation and the use of nature (Attfield 1983, Chap. 2, 154; 1999, 39–40; see Chap. 9, in this volume).

Nonetheless, other ethics are possible. A sentience-oriented ethics uses the capacity for suffering, enjoyment, and having interests as the basis for being considered a moral entity. This approach maintains that a non-human life deserves moral consideration in itself and not merely as a function of satisfying human needs (Singer 1977, 8–9, 215). An animal rights ethics is grounded on the capacity to be an experiential subject-of-a-life, a capacity shared by both humans and animals, and considers both to be equally endowed with inherent value and moral rights (Regan 1983, 235–250; 1985, 14, 21–23; see Chap. 8 by Regan, in this volume). There is also a biocentric ethics, where animal and plants – with a good of their own that can be promoted or damaged by human moral agents – are considered as bearers of moral and legal rights (Taylor 1986, 222).

There are even some "ethics of the unthinkable" – those which overcome the traditional anthropocentric ethical standards. Such ecological ethics hold that all the biotic and abiotic components of the natural environment have moral and legal rights on their own account (Stone 1972, 456, 501), and that humans have the duty of protecting the integrity of species *in* the ecosystems, and defending the forms of life that generate and sustain the ecosystems of which they are a part and an expression (Rolston 1985, 724; 1988, Chaps. 4 and 5). Aldo Leopold (1887–1948), an ecologically and philosophically enlightened forester, was the forerunner who opened the door to this kind of "extension of ethics" that involves stepping beyond the traditional instrumental relationships that humans have with the rest of nature. With an accent reminiscent of Thomas Henry Huxley's (1825–1895) worldview (see below), Leopold maintained that if ethics, basically, prescribes conducts that restrict freedom of action in the struggle for existence, thanks to which groups evolve towards cooperation, humanity now needs to recognize itself as part of a larger moral community involving both the biotic and abiotic components of natural systems. He claimed that an ethics of respect for non-human components of the ecosystem is the necessary product of social evolution. This would mean a society where humanity no longer considers itself as the "conquerors" of nature, but as "members" of a an enlarged community, where economics no longer determined all the uses of ecological systems that are uncritically considered as simple "natural resources" (1966, 217–220 [1949]). For this kind of non-anthropocentric ethics, the persistence of the intrinsically valuable biosphere, with its integrity, complexity and dynamic stability, depends on the preservation of species and ecosystems. (Callicott 1989, 142).

1.2 Looking for an Epistemic and Practical Meta-paradigm: The Transactional Framework

In science as in the rest of life, everybody wants to rule the world. But the specific beauty of science is that at one moment the ideal dimension, the theoretical systems, cease to be autonomous and unrelated to empirical reality, at one moment they cope with the reality of experience. A transactional worldview is not necessarily or intrinsically more representative of ontological reality than other views. But it can be an epistemological and methodological tool that allows us to take into consideration some aspects of reality that previous models have not been able to take on board.

Scientific knowledge is structurally based on the use of epistemic fictions that allow us to come nearer to ontological reality, whatever that reality is. But some of them are both more fruitful and more corroborated than others, or rather, they will potentially help us more than others to discover new aspects of this reality. In other words, if the process of scientific knowledge is a kind of asymptotic process, in the sense that the sciences continuously approach ontological reality, up to infinity, without ever fully reaching it, some of these theoretical tools allow us to grasp specific aspects of the reality better than others do.

This paper places an epistemological wager on the transactional worldview, which is considered as one of these useful fictions that can help us to deal with some convergent aspects underlying the research in ecology, evolutionary studies and moral philosophy. What is clearly emerging from these studies is that the environment is acquiring new senses and values. First, the dichotomy between organisms and the environment is tending to disappear. Second, some elements emerging from the analysis of ecological, evolutionary and moral studies are converging with respect to the processes of co-determination between organisms and the environment. The environment is ceasing to be a simple "filter" or "background" to biotic dynamics and becoming a real, concrete protagonist on the ecological, evolutionary and ethical scenes.

John Dewey (1859–1952), in collaboration with Arthur Fisher Bentley (1870–1957), wrote a book, *Knowing and the Known* (1949), in which some of the foundational ideas that he had already sketched out in previous books and papers reached maturity. Key to this was a historical-evolutionary analysis of the forms that have characterized scientific inquiry and correlated types of knowledge.

Three procedures or levels of inquiry historically follow one another, based on "self-action," "inter-action," and "trans-action"[1]. A self-actional procedure considers things as possessing powers of their own and as acting under their own power (1989, 66, 101 [1949]). More precisely, self-action is a "Pre-scientific presentation in terms of presumptively independent 'actors,' 'souls,' 'minds,' 'selves,' 'powers' or 'forces,' taken as activating events" (*Ibid.* 71). An interactional procedure, instead, happens where thing is balanced against thing in causal interconnection (*Ibid.* 101), where there is a presentation of particles or objects organized as operating upon one another (*Ibid.* 71). Dewey and Bentley eliminate any ambiguities between "inter" and "trans," confining the prefix "inter" where "in between" is dominant, and employing the prefix "trans" where the mutual and reciprocal are intended (*Ibid.* 264–265; see also Ratner and Altman 1964, 125). The transactional level of inquiry occurs:

> . . . where systems of description and naming are employed to deal with aspects and phases of action, without final attribution to "elements" or other presumptively detachable or independent "entities," "essences," or "realities," and without isolation of presumptively detachable "relations" from such detachable "elements" (Dewey and Bentley 1989, 101–102).

A transactional perspective is used when it is not possible to describe either component of the system adequately without implying the others (Ratner and Altman 1964, 301). The transactional formulation can be a useful method of posing and analyzing problems in quantum physics (1989, 107–109; Ratner and Altman 1964, 631), and in embryological, evolutionary and ecological domains where the historical component is prevalent and where such a procedure becomes an epistemic necessity (1989, 97, 120; Ratner and Altman 1964, 527).

Therefore, from a transactional viewpoint, any observation of totalities, parts, elements and relations is nothing but a methodological abstraction. In other words, the transactional approach adopts as a reference entity the "whole" of the events – including the relation between the knower and the known – without identifying the eventual entities and the surrounding environment as *substantiae*, i.e., things that are ontologically separated and subsequently are found to have relationships. At the same time, a structural transactional network or "web" is presupposed to be the "logical" primary reference (for the relation between ontology and logic, see: Dewey and Bentley 1989, 287).

Looking through the tangled history of ecological, evolutionary and ethical studies, it is possible to catch a glimpse of an ever-lasting underlying tendency. These scientific disciplines are crisscrossed by an integrative, inclusive, monistic and systemic tendency towards the complete overcoming of the dichotomy between organism and environment. Once we become aware of the existence of such a convergence, it is possible to establish a meta-paradigm, a connecting epistemological framework built on this common orientation. Furthermore, a transactional framework is better adapted than other frameworks to convey this common, shared epistemological ground.

In fact, assuming a transactional framework, the following positions, among others, acquire a clearer and more univocal meaning. Tansley's concept of the ecosystem definitively settled the split between the biotic and abiotic environments and between the different biotic compartments (plant and animal communities). Odum's systems ecology was proposed as a new approach that apprehended the specificities of the ecosystem as an emergent whole that was not reducible to its biotic and abiotic components. Andrewartha and Birch explained population dynamics based on the reciprocal action and reaction between organisms, constituting local populations, and the environment. Oyama's developmental system theory claimed that non-genetic factors, both biotic and abiotic environmental factors, participate causally in ontogenetic development and evolution. Genes as prime movers of evolutionary processes cease to exist. What exists, and what has developmental and evolutionary significance, is the organism-environment system, with its many levels of organization and causality. Biological and behavioural structures and functions of the organism result from the developmental and evolutionary interdependence of organism and environment. Finally, in moral philosophy, Leopold, Rolston and Callicott proposed a new environmental ethics grounded on the extension of the moral community to non-human living and non-living components of the ecosystem.

Ultimately, with regard more specifically to the evolutionary explicative concepts of "replicator" and "interactor," it is worth noticing that using the epistemological trilogy of Dewey and Bentley, Dawkins's replicator proposition can be interpreted as pre-scientific, i.e., the self-actional character attributed to the genes is an element in constructing a fiction that passes over all the gene-complexes, developmental and environmental determinisms (see also the anticipatory critique by Dewey and Bentley of the gene concept, 1989, 118–119; Morange 2001, 88–90, 159–163). David Hull recognizes that "... the distinction between an organism and its environment is ... artificial" (1979, 429; see also Sober 1984, 87). In any case, in the development of his interactor concept this distinction is a central assumption. The interactor is an "individual" – the genes, and all the remaining levels of organization, with the exception of the communities and ecosystems – that interacts with the environment, understood as structurally separated and external with respect to it: "Genes of course, can also function as interactors. They interact directly with their cellular environment, but they interact only indirectly with more inclusive environments via the interactors of which they are part" (Hull 1981, 34; but see also El-Hani 2007; Meyer et al. 2011).

As a complement to the replicator and interactor concepts in an evolutionary and ecological context, the "transactor" is proposed here with the aim of taking into consideration the permanent, mutual, reciprocal relationships between the environment and the entity under selection. In contrast to an interactional framework, where the environment/entity are viewed as in a causal relationship, but as definitely separate. Such a separation ordinarily confers causal preeminence on the inward biological factors over the environmental ones (Bergandi 2007):

- The transactor identifies a functionally cohesive, coherent, complex, and relatively independent (or autonomous), environmental-organic entity[2];
- The transactor is a methodological construct that integrates into the definition of an evolutionary entity those environmental factors that have selective value for its existence;
- The transactor is part of a transactional web with other entities of similar, lower and higher levels of complexity;

- The identification of such an entity implies the attribution of specific emergent properties that may express specific adaptations or ecological properties;
- The need to take into consideration the upper transactor is revealed when the differential frequency of the proliferation of an entity (gene, organism, deme, population, species) is sensitive to, or depends on, its "context."

In an evolutionary perspective, at least from the gene level, any level can be considered as an environmental-organic or transactional totality. The transactor, indeed, integrates into the definition of an entity those environmental factors that have selective value for its existence. This concerns what Brandon identifies as the "ecological environment," and the "selective environment" (1988, 57; 1990, 47, 49; 1992, 81–86). The theoretical core of this transactional perspective is the idea that there is a permanent functional connection "in" the transactor between the entity selected and the ecological-selective environment.

For instance, the genome is the immediate environment of the single gene, or of a complex of genes, and the ontogenetic and evolutionary values of a gene are determined to be an integral part of such an entity. Similarly, as far as the other organizational levels are concerned, they will be organic-environmental entities, like transactors. The biotic and abiotic components of ecosystems that have selective values for the transactor in question must be considered parts of the evolutionary and ecological connotations of the individual organisms, the populations, and the species.

Finally, a direct and fundamental consequence of this methodological proposition is support for a legitimate explicative pluralism, until proved otherwise (see Sober and Wilson 1998; Okasha 2006). *A priori* every transactor, or transactional level, legitimately has a possible causal role in the determination of adaptations and other evolutionary, and ecological, processes that can be generated in the whole range of biological organizational levels. On the other hand, ecology can be perceived as the scientific transactional discipline par excellence. From population and community ecology to ecosystem ecology the theoretical and experimental models grasp at different levels the intrinsic interdependence of biotic and abiotic components of the ecological systems. Such a transactional framework could play a structuring role in conservation biology, where the need for interdependence is extended beyond natural systems so as to integrate human society as one major component of what is becoming the "planetary socio-ecosystem."

In a moral philosophy context, if the experimental method were adopted in the conduct of ethical and social affairs, as Dewey reminds us, a transactional perspective would be one of the possible ontological frameworks. It would not be taken as rigidly established prior to and independent of scientific inquiry, but rather as a hypothesis, the consequences of which should be tested. In fact, it could be one possible basis for an extension of ethics, for a more integrative moral ontology, i.e., the enlargement of the moral community to the biotic and abiotic components of the ecological systems.

Once the experimental method in ethics was adopted, principles, rules and beliefs "...would be recognized to be hypotheses. Instead of being rigidly fixed, they would be treated as intellectual instruments to be tested and confirmed – and altered – through consequences effected by acting upon them. They would lose all pretence of finality – the ulterior source of dogmatism" (Dewey 1984, 221 [1929]; for a similar position on the development of environmental ethics and the necessity of an environmental pragmatism, see: Norton 2005, 2007; see Chap. 6 by Blandin, in this volume). This link between the experimental method and ethics would open up possibilities for setting a new epistemological course based on the merger of ecological, evolutionary, and ethical studies and issues, in a virtuous epistemic and practical circle.

1.3 Evolution between Ethics and Creationism

The Darwinian evolutionary paradigm has not only revolutionized the biological sciences, having become the explicative epistemological background for all biological phenomena, but it has also had obvious consequences on ethical and social constructs.

Thomas Henry Huxley (1920 [1894]), the most influential supporter of Darwin, recognized that human social life is both a part and a product of the cosmic process determined by the struggle for existence, i.e., by the selection and survival of those forms of life that are best adapted to the environmental conditions. At the same time, Huxley's reading of the evolutionary process, which was largely influenced by Herbert Spencer (1820–1903) (see Chap. 2 by Ruse, in this volume), implies that a "progressive" development from uniformity to complexity has been the deep rule driving the occurrence of natural events. In Huxley's view, the notion of progressive development applies not only to the history of the cosmic process, but also to human social contexts. In the highest and most complex stages of social development, the emergence of cooperative behaviors gains the upper hand over the struggle for existence, which is the agent of the selective process in the state of nature. The fundamental reason advanced by Huxley to explain this inescapable feature of social development is the following: the progressive limitation of the struggle for existence between the members of society leads to increasing efficiency as regards outside competition, either with the state of nature, or with the members of other societies. This would be the only relational context that allows for the preservation of the bonds that hold members of a society together (1920, 34–37, 51–53, 79–83). Finally, in human society the "ethics of evolution" implies a distancing from instinctual, compulsory self-assertion and an embracing of self-restraint. That is, it involves repudiating a gladiatorial vision of existence, so as to escape from the animal kingdom and establish a kingdom of man ruled by the principle of moral evolution, where "social life is embodied morality" (Huxley 1907, CCLXXV [C.E., ix, 204]).

It is definitely enriching to look back at John Dewey's criticism of Huxley's *Romanes Lectures* on evolution and ethics, where he points out Huxley's oxymoronic structure of thinking (1898). According to Dewey, there is no reason to oppose the natural process and the ethical process on the basis of a supposed identity between the struggle for existence and self-assertion and an arbitrary attribution to the latter of unscrupulous, gladiatorial traits. Some positive behavioural traits such as benevolence, self-sacrifice and cooperation can also be considered as part of self-affirmation and, therefore, of the struggle for existence. The Deweyan unified vision of biological and ethical evolution is grounded on the refusal of the Huxleyan dualism between cosmic and ethical processes (on the continuity between ethical and evolutionary processes, see Ruse 2009, xxiv–xxvii), and on the idea that natural selection is still operating in the same way in human social life. The main differences between humans and non-humans have to do with the fact that the range of uses for the environment is wider in human societies than for other species, and the selected functions differ: to be fit among animals does not mean the same thing as to be fit among humans (Dewey 1898, 41, 45–49, 52–53, see also: Kropotkin, below). Dewey's position finds some meaningful echoes in the way that the present-day defense of evolutionary ethics looks to the biological origins and basis of human morality: "Our moral sense is an adaptation helping us in the struggle for existence and reproduction, not less than hands and eyes, teeth and feet. It is a cost-effective way of getting us to cooperate, which avoids both the pitfalls of blind action and the expense of a super-brain of pure rationality" (Ruse 1995, 97).

Certainly, the meeting between evolutionary thinking and ethics has generated antithetical interpretations that navigate perilously between the Scylla of dogmatic religion-based morals, emanating from an intangible divine power, and the Charybdis of the naturalistic fallacy, which implies the refutation of any inference of moral rules from propositions about natural occurrences (Moore 1903). In fact, the naturalistic struggle for life that once was applied to human society took the form of an ethics of "rational egoism," whereby society as a whole was held to benefit from the competition among individuals struggling for the acquisition of the means of subsistence (Spencer 1892, 1st vol., 199). On the other hand, Peter Kropotkin (1842–1921), in *Mutual Aid* (1972 [1902]), proposed a more cooperative interpretation of the relationships between individuals and groups in nature. The direct consequence of such a conception is an ethical worldview based on cooperation, which shares some similarities with Thomas Henry Huxley's position – a worldview grounded on the idea that in human society, as well as in nature, "the fittest" are not those who are continuously fighting each other, but those who support one another. Mutual aid and mutual struggle are laws of Nature, but mutual support would be a major factor in evolution because it allows individuals to flourish, to rear their progeny and to develop to the best of their potential. While under certain circumstances, individual qualities such as force, swiftness and cunning certainly allowed individuals to be fittest, nevertheless " … under *any* circumstances sociability is the greatest advantage in the struggle for life" (Kropotkin 1972, 68; italics in original). Herbert Spencer, Thomas Henry Huxley, and Peter Kropotkin ethicized evolution, grounding it on the implicit idea that the underlying deep structure, sense and purpose of natural, and social, reality was an enhancing progressive tendency towards complexification and cooperation.

Here are presented some aspects of the scientific and philosophical background that underlies Michael Ruse's chapter, "Evolution versus Creation: A Sibling Rivalry?". Ruse provides points of reference for understanding the conflictual relationships between evolutionary biology and the religious creationist worldview. Basically, evolutionary studies from Darwin until today have proposed an approach to understanding the past and present and the variety of all living beings on Earth: in the great and complex battle for life, organisms result from a long, slow and gradual natural process in which the primary causative factor is natural selection based on a struggle for existence. From the beginning, such a scientific construct inevitably challenged the world depicted by religious thinking. Nevertheless, according to Ruse, the Darwinian scientific paradigm has been hijacked and twisted by its epigones, particularly Thomas Henry Huxley. In the hands of Huxley, what was a scientific approach to nature became a secular religion without revelation. Another protagonist of this hijacking has been Herbert Spencer. The idea that significantly structured Spencer's metaphysical conception of evolution as the progressive complexification of all natural and social processes has been the keystone of a major misrepresentation of Darwinism. Even recently, Edward O. Wilson metamorphosed evolutionism into the cornerstone of a scientific materialist worldview of nature and society. Such a distortion of evolutionary thinking takes upon itself the definitive power to explain even the reasons for the emergence of religious thinking and to deprive theology of meaning and strength. Finally, Ruse reminds us, mixing science and ideology risks becoming explosive. The outcome can be very negative for the persistence of the scientific, evolutionary view of life in schools and in the rest of society. At present, creationists can wield this misuse of evolutionism to argue that scientists are the vehicles of a "secular religion," and use that as justification to impose their will on and against science.

1.4 Chance and Time between Evolution and Ecology

"All-over progress, and particularly progress toward any goal or fixed point, can no longer be considered as characteristic of evolution or even as inherent to it" – George Gaylord Simpson (1902–1984) thus summarizes his evolutionary conception grounded on purposeless, materialistic and random processes (1949, 343). Progression, in the sense of a succession of phases of a process, and not progress, certainly exists in evolution, but the occurrence of environmental and genetic events based on chance and randomness makes any kind of oriented, progressive evolutionary change impossible. Such a class of events introduces us to an epistemological domain where determinism and the causality principle are confronted by chance and random processes that have causes, but that remain substantially unpredictable, or that can be treated only stochastically. For Darwin, variations were due to chance, the causes were unknown, but they were natural causes and did not require any mysterious source (on the role and the meaning of chance in Darwin, see Morgan 1910; Ruse 2008).

Does chance exist? "Though there be no such thing as *Chance* in the world, our ignorance of the real cause of any event has the same influence on understanding, and begets a like species of belief or opinion" (Hume 2007, VI, 41, 1 [1748]; italics in original). Is the term 'chance' merely a negative word, veiling our ignorance of the real causes underlying the occurrence of the phenomena, and leveling the specific contribution of the various events contributing to the occurrence (*Ibid.* VIII, 69, 25)? What is the relationship between cause, purpose and chance? Is chance a relative notion that is contrary to purpose, but not to cause? (see: Katz 1944). In evolutionary studies, what role is there for: (a) the statistical meaning of chance, resulting from the confluence of an ignorance-based interpretation of chance (dating back to Hume and Laplace) and the idea that a chance event is determined by the intersection of independent causal chains of events (for the latter conception of chance, see Chap. 3 by Gayon, in this volume, and Warren 1916), and (b) the evolutionary meaning of chance, arising from the idea that events are independent of an organism's needs and of any directionality provided by natural selection in adaptive processes (Eble 1999)?

To begin to answer these questions, it is necessary to identify several of the partially overlapping meanings that are attributed to the word 'chance' in evolutionary studies. Jean Gayon, in his analytical essay "Chance and Evolution," identifies these meanings (luck, randomness and contingency) as well as the contexts (mutations, random genetic drift, genetic revolutions, ecosystems and macro-evolutionary events) in which the word 'chance' appears.

The author reminds us that when genetic mutations emerge randomly, "by chance," biologists tell us that they are "fortuitous" or "unexpected." One could say that advantageous mutations are like jewels that a gardener "chances" to find while working in his garden. In the case of random genetic drift, the process is random in the sense that it allows certain alleles to fix stochastically in a specific locus in a Mendelian population, following the laws of probability. As in roulette or a game of dice, we do not know precisely the initial conditions, and we can say only that there are several possibilities, without being able to predict precisely which allele will ultimately be fixed.

At the level of the genome, genetic drift in isolated populations contributes to determining "genetic revolutions" that involve a different kind of chance, one that directly concerns the theoretical system. As a consequence of the complexity of the interactions between genes – and with the environment – genetic revolutions are fortuitous in the sense that they are not predictable for contingent reasons. Gayon points out that these reasons can be: (a) it is not within the range of the theory to predict such events; (b) the initial conditions are not sufficiently known; or (c) the complexity of the phenomena precludes prediction. This last reason is also the cause of the unpredictability of fortuitous interactions (between organisms, and between organisms and the environment) peculiar to the ecosystem level.

At the macro-evolutionary level, when the "contingency" of evolution is addressed, two meanings of "chance" are put forward. Firstly, due to the contingency and complexity of the history of life, our paleobiological theories will never be able to retrace exactly what has happened. Secondly, the survival or extinction of some species is contingent because of accidental occurrences (such as a swift, drastic change of environmental conditions, for example) and not because of their specific capacity for adaptation. Finally, Gayon ends his essay with a cross-reference to Antoine-Augustin Cournot (1801–1877, 1843), who proposed a specific sense of chance that could be of use in evolutionary studies.

Chance, in its various forms, contributes to shaping evolutionary and ecological processes at many levels of organization, from genes to ecosystems, according to evolutionary and ecological time scales. Evolutionary time deals with time scales that allow for gene frequency changes in populations that can lead to speciation and evolutionary adaptation. Ecological time deals instead with a shorter time span and concerns essentially species dispersal and the complex web of relationships that populations establish with their immediate environmental factors (Pianka 2005). Evolutionary and ecological systems are definitely and inextricably intertwined, as are evolutionary and ecological times. The study of the ecological outcomes of evolution – the properties of biotic populations and communities resulting from the natural selection of their components – and the speciation processes ensuing from ecological pressure over geological time, involves taking into consideration these two types of time span. It is noteworthy that, in both evolutionary and ecological contexts, there is room for another type of time, totally different from the classical chronological time, which follows a sequential series of moments. This other time is kairological (from καιρός: *kairos*; the Greek god of a non-sequential, non-chronological time: a propitious, opportune, right time). This is a time of contingency, a time of right or adapted behavior, at the proper time involving, in the struggle for life, the difference between the life and death of the organisms, with all the ensuing consequences for the evolution of the populations (for an analysis of kairological time and contingency, see respectively Gault 1995; Gould 1999).

In his chapter "Some Conceptions of Time in Ecology," Jean-Marc Drouin analyzes the succession of different concepts of time that have characterized the development of ecology from the end of the eighteenth century until modern times. Geology dismissed the short biblical chronology and made it possible for a long-term history of the Earth to become the scientific basis for all natural sciences. Ecology, while a historical science like geology, is also a science of processes, like physiology. This specificity determines some of its peculiar characteristics. Fundamentally, ecological processes have been described according to three different paradigms based on the ideas of cycle, growth, and chaos, the latter entailing unpredictability. The different conceptions of time underlying these paradigms can sometimes be intertwined, as in the case of climactic conceptions of plant ecology (cycle and growth) or mathematical models of population ecology (cycle and chaos).

A cyclical conception of time, already present in botanical geography (Alexander von Humboldt (1769–1859)), was adopted by the pioneers of plant ecology. Henry Chandler Cowles (1869–1939) and Frederick Clements, using the notion of "climax," proposed a conception of the succession of vegetational stages over time that culminated with a "peak" specific to a given geographic region. In this case, the ideas of cycle and growth coexist: a forest, as an organism, develops and dies, always following the same successional structure. Later, the works of Lotka, Volterra, and Umberto d'Ancona contributed to applying mathematical models to the prediction of cyclical fluctuations in prey and predator populations, which are dephased over time. Finally, Drouin reminds us, Robert May has shown, under some circumstances the dynamics of populations can be unpredictable over time. In the case of populations where growth takes place at discrete intervals, as in some insect species with non-overlapping generations, and where rates of growth are highly variable, the dynamics of these populations are chaotic and, hence, unpredictable.

1.5 Ethics between Ecology and Evolution

The Darwinian theory of evolution and the emergence of an ecological, scientific way of thinking are the latest stages in a gradual process of our understanding of how humanity is integrated into nature. Some previous steps in this process included, during the Renaissance, the recuperation of the naturalistic knowledge of the ancient Greeks and Romans, and the Copernican revolution. The new world-views emerging from these developments have decentralized man's place in the universe. Ethically speaking, however, the process of man's integration in nature, with the concomitant changes in perception and practical conduct, is relatively recent. The Kantian moral perspective of the *Foundations of the Metaphysics of Morals* attributed to any entity not provided with rationality a relative value as means: therefore all sorts of natural non-rational entities are, uniquely and unambiguously, means. The self-evidence of the ethical axiom according to which rational beings exist as ends in themselves supported an anthropocentric, coherent and legitimate moral system. In the philosophical backyard of the Universal Declaration of Human Rights, in good company with Aristotle and Locke, Kant played a fundamental role in maintaining that rational human beings were morally autonomous and equal. Nowadays, our ethical assumptions are still embedded in a moral world structured around this postulate (Kant 1990, 45–46 [1785]; see also: 1963, 120 [1775–1780]).

The way relationships between humans and their natural environment are modeled gives practical expression to a system of values and beliefs. Nevertheless, we must be reminded that value systems as well as species are selected by the environment. And, in our times, the global environmental crisis has helped to push us to reassess the ethical foundations of our societies. In fact, some environmental indicators are telling us that our current value system is most likely no longer adequate to deal with economic globalization and its environmental challenges. Moral adjustments are necessary as we become increasingly aware that we are living on a kind of "spaceship" where our reservoirs of resources are not unlimited (Boulding 1966, 9–10). Humans need to become aware that, as Walter Penn Taylor (1888–1972) affirmed in 1936 – referring to Henry Agard Wallace (1988–1965), who had pointed out the need for a Declaration of Interdependence to resolve political and economical international problems – "There is ... equal need for a declaration of interdependence among plants, animals, and their environment. Such a declaration may well be the preamble of the ecological constitution" (335).

Such an ecological constitution would set out that humans solemnly recognize being part of the same co-evolutionary system as other species, all of which must be preserved and respected.

As a matter of fact, the human species and the environment subsist in a co-evolutionary relationship. So Bryan Norton reminds us in the chapter where he explores the potential for a system of adaptive management of the environmental crisis built upon the Darwinian selectionist paradigm. This kind of approach, combining an evolutionary perspective, experimentalism and pragmatism, has the merit of overcoming the classic intrinsic value issue, with its constantly shifting borders of moral considerability. In this case, the core of the issue no longer concerns the entities that are acknowledged as having intrinsic value (humans, non-human world), but rather the relationships between descriptive and prescriptive assertions, "facts" and "values," environmental sciences and environmental ethics.

The foundational background of this Darwinian adaptive management is a commitment to naturalism, linked to a social context that provides for experimental confrontation of the beliefs and values involved in a given environmental problem. This enables a gradual re-configuration of the perception of the factual elements that constitute the environmental problem. In this way, a new conservationist consciousness emerges and the spatio-temporal model of the situation is broadened with respect to both natural and sociological entities and processes. Norton applies this approach to the way in which the residents of the Chesapeake Bay in the US became aware of the causes of the pollution in their area, and shows that only by taking into consideration a space-time model on a larger scale was it possible to change the ethical commitments of the social communities involved. In fact, the perception and the policy actions of the Bay residents changed only when they radically transformed their worldview (scientific and ethical). A change in the model used to think about the problem of pollution in the Bay led to a new watershed system model, which largely surpassed the geographical limits of the Bay, thus involving neighboring regions and establishing active cross-state cooperation.

Furthermore, the type of management proposed in Norton's chapter is "adaptive" and "Darwinian." The epistemological ground is a stripped-down version of natural selection: a source of variations in a population, a means of coding, and the survival of variation in the population. In the context of an environmental crisis, a community or culture will survive only if it plays an adaptive game with regard to its environment. On the one hand, individuals must survive to reproduce and contribute to the gene pool. On the other hand, the group or community must "select" and accept a sense of responsibility and stewardship for resources and the integrity of natural systems. This is achieved by setting aside forms of individual or group wellbeing that are grounded in short-term economic choices. Only in such a way will a community – whose goal is multigenerational sustainability – be able to survive, proliferate and develop, preserving a viable range of choices for future generations. A sustainable society structured around this type of selective adaptive management would be a clear expression of a concrete and fruitful meeting of ethics and the evolutionary paradigm.

Evolutionary ethics considers the possibility that moral norms contribute to humanity's success within the biosphere, and that at least the capacity to behave ethically should have been shaped by evolutionary processes. In his chapter, Patrick Blandin explores the hypothesis that some environmental ethics views, inspired by ecological knowledge, are attempts to increase mankind's adaptability. He first recalls some fundamental points in the history of ecology and nature conservation. Clearly, from the end of the nineteenth century to the second part of the twentieth, ecology constructed a view of the natural world as an "equilibrium world" (disturbed only by human activity), a process culminating with the elaboration of the ecosystem paradigm. During the first part of the twentieth Century, ecology did not have a strong influence on ideas about nature conservation, but it played a very important role in Aldo Leopold's way of thinking. Leopold's Land Ethic was explicitly inspired by current ecological ideas, and his ideal was to preserve the stability of natural communities. Equivalent ideas were at the core of the conference organized in 1949 by the International Union for the Protection of Nature, which had been created in 1948. There the "balance of nature" idea played a paradigmatic role, later reinforced by the development of ecosystem cybernetics, which supported the idea that ecosystems are normally in a dynamic equilibrium state, allowing a cyclical functioning. Furthermore, Darwin's theory has been used by ecologists to support the idea that natural communities reach an equilibrium through natural selection: evolutionary change should produce ecological stability. But, during the last decades, an important paradigm shift has been taking place. Ecological systems appear to be changing through time, along trajectories that are unique and chaotic, and thus unpredictable, even if the processes are deterministic: ecology meets evolution, understood as a global process of change. Blandin proposes viewing the biosphere as a "transactional web," (see previously 1.2) where interacting entities are "co-changing": a "co-change paradigm" is taking the place of the "ecosystem equilibrium paradigm." Consequently, there is a need for an eco-evolutionary ethics, as a new step in the development of environmental ethics. Recalling Julian Sorell Huxley (1887–1975), Otto Herzberg Frankel (1900–1998) and Michael Soulé, who advocated human responsibility for the future of evolution, Blandin argues that the aim of conservation is to maintain the biosphere's adaptability. He therefore focuses on the problem of biodiversity conservation, raising thorny questions about the uniqueness of living entities – which is connected with the intrinsic value issue – the evolutionary meaning of diversity, and the substitutability between species presumed to be functionally redundant. An "eco-evo-ethics" must assume that we are living in a changing, transactional web, and provide relevant principles and guidelines. But, at the same time, it must assume that values may also change, through permanent transactions between eco-evolutionary science and environmental ethics.

Moreover, ecology can serve as a metaphorical ground for new moral forms that allow more integrative, ecological ethical conducts. The contribution by J. Baird Callicott, "Ecology and Moral Ontology," analyzes the important role played, during several decades, by the organicistic paradigm in the history of ecology. According to Callicott, pre-energetic Clementsian ecology is more indebted to the biological conceptions of his time than to extra-scientific philosophy and sociology (in particular, German idealism and Durkheim's sociological functionalism). Clements, via analogical reasoning, structured ecology along similar lines with biology, merging ecology with physiology and looking for precise knowledge about the conditions of the life of plants, about the external factors in the environment in which the plants live and about the activities that these factors determine (Clements 1907, 1). This eco-physiological approach replaced the study of the functional interrelations of organs in organismal biology with the study of the functional interrelations of species in organicistic ecology. Likewise, the study of organismal development, in the context of Clementsian ecology, was replaced by the study of the successional development of the plant formation. With the arrival of the ecosystem paradigm, the organicistic framework underpinning ecology up to then gradually but inexorably faded from the scene. Some specific characteristics of the natural systems clearly emerged: successions are not directional and predictable, species populations are not in a steady-state equilibrium and ecosystems are not superorganisms.

Callicott prefers to reverse the metaphor, recognizing the organisms as "superecosystems," in the sense that the organisms "magnify" the characteristics of the ecosystems, and exhibit in a superior way the characteristics attributed to the ecosystems: the organisms as the ecosystems are hierarchical, constituted by many different types of subsystems, self-regulating and open to environmental energies and relationships. For example, the author reminds us, the metabolic processes of the organisms are carried out by a multiform variety of species populations of the internal microbial community (on this topic see also: Palka-Santini and Palka 1997).

From a moral point of view, the modern, traditional, Cartesian moral ontology, grounded on the dichotomy between subject and object, between thinking and not thinking entities, considers that moral thinking monads could have moral relationships only with other entities possessing the same moral essence. According to Callicott, on the contrary, the organism-as-superecosystem metaphor represents the core knowledge concept needed to construct a post-modern ecological moral ontology that departs significantly from the Cartesian perspective. This core allows the recognition of the ontological continuity between our own self and our biotic, and abiotic, surroundings. One's self is a "nexus" in a flux of relationships connecting internal and external organisms to one's superecosystem. And above all, our existence as nexus, and not as monadic self, allows us to imagine, and practice, a very nuanced, hierarchical system of ethics based on the proximity (cultural, historical, geographical) with people, institutions, things and environments, an "inclusive ethics of care and concern for those people, institutions, places, and things that define oneself and give meaning to one's life."

At present, the phylogenetic continuity between man and other living beings maintained by evolutionary studies and the inescapable structural embeddedeness of human species in ecological systems have been clearly integrated into more recent moral philosophy, promoting an enlargement of the boundaries of the ethical community. However, even at the end of the nineteenth century, Henry Stephens Salt (1851–1939), who already supported the recognition of animal rights against human supremacy, was directly referring to the ethical arguments of Jeremy Bentham (1748–1832, 1789) and John Stuart Mill (1806–1873, 1848) against cruelty to animals. Salt argued that modern science recognized as a fact that between man and the other animals there was only a difference of degree and not a difference of kind, and on that basis proposed an enlargement of the moral community based on the extension of the idea of "humanity" to the other animals. This philosophy of rights pushed for an ethical reform grounded in "the recognition of the rights of animals, as of men, to be exempt from any unnecessary suffering or serfdom, the right to live a natural life of 'restricted freedom,' subject to the real, not supposed or pretended, requirements of the general community" (1894, 85). Salt reminds us that in the past such an extension was made to "savages" and slaves, and that if humanism does not wish to become divorced from humaneness, it must embrace non-human animals.

Tom Regan's contribution fits into this heterodox tradition of moral philosophy. Regan's chapter expounds on the defining characteristics of his "rights view," which he considers the most appropriate ethical position vis-à-vis the global environmental crisis. Basically, for individuals to possess moral rights means: (a) that others are not free to harm their life, body and liberty; (b) that these rights are possessed equally by all; and (c) that respect for these rights is the foundational meta-ethical grounds of all morality. To be the "subject-of-a-life" is the main requirement for the possession of these rights. Nevertheless, such a characteristic does not limit these rights to humans. Other animals (mammals, birds, maybe even fish) possess them because, like humans, they are in a world of which they are aware and they are concerned with what happens to them. In other words, humans and other animals share a similar kind of subjectivity. This ethical framework has revolutionary implications: the end of all commercial use of animals for food and of the human predation that we call the "sport of hunting."

Regan also refutes environmentally-based objections that the rights view necessarily involves the extermination of predators, since predators violate the rights of their victims (Callicott 1989), and that it fails to provide a consistent basis for the preservation of endangered wild species. Firstly, in terms of predator–prey relationships, he defends the natural *laissez-faire* brought about by the struggle for existence. Secondly, he supports the applicability of some compensatory principles of justice to preserve species endangered as a consequence of human action. These species would have the right to compensatory assistance: in other words, humans owe them assistance because of the selective disadvantage that they caused them. At the same time, however, as a direct consequence of his ethical assumptions, and because of the absence of definitive convincing arguments to the contrary, Regan denies that species, ecosystems or the biosphere have intrinsic value.

Observing current developments in moral philosophy, it is not excessive to say that there are many different environmental ethics, almost as many as there are philosophers supporting them. However that may be, it is possible to identify some shared ontological and ethical foundations that make it possible to distinguish a minimal common basis for environmental ethics:

- Humans, like other species, are members of the Earth's single biotic community;
- Humans, like other species, are an integral part of a system of evolutionary and ecological, biotic and abiotic relationships that allows them to survive, proliferate, and develop to the best of their potentialities;

- Humans must control their proliferation and their economic development in a way that allows for the highest possible level of biodiversity and evolutionary potential on the planet.

Concerning other foundational topics there is no unanimity. Moral philosophers proceed in a scattered order, supporting positions that are frequently mutually exclusive. One of these topics is the issue of intrinsic or, depending on the author, inherent value. We have already seen some elements of this philosophical debate about a moral ontology, about the entities accorded moral standing or the functions that make possible the attribution of such a standing. What entities possess intrinsic value? Only humans (Norton 1987, 2003, 2005)? All living entities equally (Taylor 1986)? All living beings, ranked by degree of intrinsic value (Rolston 1988)? Species, biotic communities and ecosystems (Callicott 1986, 1989; Rolston 1988)? All the individuals that are subjects-of-a-life – having perceptions, preferences, desires, etc. (Regan 1983, 1985)? Or individual living organisms that have a "good" of their own (Attfield 1987, 1999)?

With this intrinsic value issue as an ontological background, Robin Attfield's chapter explores one of the recurrent issues in environmental ethics: can environmental ethics reconcile individualism with a more ecological, holistic perspective? Is monism or pluralism the best meta-ethical framework to achieve such a goal? Attfield takes Carter's review of *The Ethics of the Global Environment* (2001) as a starting point for analysis. In this review, Carter asserted that theories of value-pluralism are better adapted than monistic theories to cope with deeper ecological values, such as species and population preservation. Value-pluralism critiques the maximization of one value at the expense of other values, refutes comparability among values and maintains the incorporation of priority-values that specifically characterize normative theories into a broader theory. Attfield's critique of this pluralistic view is irreproachable. Such a combination of values is considered a source of contradiction, because anthropocentric, zoocentric, biocentric and ecocentric values are incompatible and mutually exclusive. Moreover, any decision-making process involving such antithetical values would lead to complex and litigious policy choices. Finally, the most recurrent monistic theories, even if they hold to a specific value, simultaneously also honor other values. The Attfield critique simultaneously shows the inconsistency of pluralist attempts to combine values and the impossibility of reconciling ecocentrism with many forms of individualism.

Catherine Larrère's essay goes on to take into consideration the debate on the modalities of existence of intrinsic value in nature, in correlation with the analysis of an antithetical philosophy of the environment: the philosophy of technology. The philosophy of nature, arguing for intrinsic value, seeks the best ethical and meta-ethical principles to preserve nature in its many forms and levels – wilderness, wild species, ecosystems, biosphere. Such a perspective is grounded in the assumption that there is probably something wrong with the relationship between humans and nature. The philosophy of technology, on the other hand, does not consider mankind responsible for our ill-adapted relationships with the environment, but sees our incapacity to master technology as the main cause of the environmental crisis. Behind these philosophies are an ethics of respect for nature and an ethics of responsibility. Pragmatically speaking, Larrère reminds us that the preservationists would leave a wilderness alone (let a forest burn, for example), whereas the technologically-oriented conservationists would prefer to intervene in natural processes, taking into account the development of human societies (controlling forest fires to limit the greenhouse impact, for example). Larrère concludes that this confrontation between the philosophy of nature and the philosophy of technology is no longer useful for understanding and pragmatically coping with the environmental crisis because of the proliferation of "hybrid objects" that are neither totally artificial, nor totally natural (global warming, GMOs, etc.). Specifically, arguing against Latour's persistant dualism (1999) – nature versus society and technology – Larrère privileges a synthesis of an ethics of respect and an ethics of responsibility. Only such an ethical perspective would be capable of recognizing that nature is a standalone entity, and at the same time that nature is no longer a "given" but the outcome of scientific and social controversies.

A tendency towards anthropocentricity is connatural to the human species. Without this propensity we probably would not have been able to survive. In fact, during evolutionary times we have had to contend with nature to proliferate and develop our civilizations, which directly emerge from this confrontation with nature. The point of no return was reached when humanity was able to overcome its direct dependence on nature, when humanity achieved the lasting ability to adapt its environment to its needs, and not simply to follow evolutionary and ecological processes in the same way as all other species. Of course, we are embedded in nature, and we are ontologically dependent on ecological systems and evolutionary processes. But at this stage of our history we have available many different ethical options for the development of our societies and our possible relationships with nature.

The current anthropocentric and globally dominant ethical worldview emerges from this history of relationships between man and nature, and we must recognize that this helped us to find our place in the world. The crucial question at issue is this: nowadays, is anthropocentrism, even in its weakest forms, the most suitable way to cope with the environmental crisis and the decline of biodiversity, which, practically speaking, are the direct results of this ethical worldview? To identify intrinsic value only in man, or to identify a ranking of intrinsic values in living beings, expresses the traditional religiously or philosophically-grounded hierarchicalist Great Chain of Being worldview. Is it possible to reform these positions in an environmentally-oriented sense that could radically change relationships between mankind and nature? Or, on the contrary, do we require an epistemological and ethical rupture with respect to the idea that humanity has of itself and of its place in the world?

We hope that the contributions in this volume will provide some elements of a response to the complex weave of evolutionary, ecological and moral questions that are posed with respect to the possible future pathways of development of humanity's relationship with the rest of nature.

Notes

1. The word 'transaction' and some elements of the corresponding methodological procedure was already proposed in *Conduct and Experience* (1984, 220 [1930]) by Dewey, where he points out that in the complex organism-environment: "Only by analysis and selective abstraction can we differentiate the actual occurrence in two factors, one called organism and the other, environment".
2. The term 'environment' refers not only to the environment of the organism, but also to any other "context" of the entities under selection.

References

Allee, W.C., O. Park, A.E. Emerson, T. Park, and K.P. Schmidt. 1949. *Principles of animal ecology*. Philadelphia/London: W.B. Saunders Company.

Andrewartha, H.G., and L.C. Birch. 1954. *The distribution and abundance of animals*. Chicago: University of Chicago Press.

Andrewartha, H.G., and L.C. Birch. 1984. *The ecological web: More on the distribution and abundance of animals*. Chicago: University of Chicago Press.

Attfield, R. 1983. *The ethics of environmental concern*. New York: Columbia University Press.

Attfield, R. 1987. *A theory of value and obligation*. London/New York: Croom Helm.

Attfield, R. 1999. *The ethics of the global environment*. Edinburgh/West Lafayette: Edinburgh University Press/Purdue University Press.

Baxter, W.F. 1974. *People or penguins. The case for optimal pollution*. New York: Columbia University Press.

Bentham, J. 1823/1907. *An introduction to the principles of morals and legislation*. Oxford: Clarendon Press. 1st ed., published in 1789.

Bergandi, D. 1995/2000. "Reductionist holism": An oxymoron or a philosophical chimaera of E.P. Odum's systems ecology? *Ludus Vitalis* 3(5): 145–180. Repr., *The philosophy of ecology. From science to synthesis*, eds. D.R. Keller and F.B. Golley, 204–217. Athens: University of Georgia Press.

Bergandi, D. 2000. Eco-cybernetics: The ecology and cybernetics of missing emergences. *Kybernetes* 29(7/8): 928–942.

Bergandi, D. 2007. Niveaux d'organisation: évolution, écologie et transaction. In *Le tout et les parties dans les systèmes naturels*, ed. T. Martin, 47–55. Paris: Vuibert.

Bergandi, D. 2011. Multifaceted ecology between organicism, emergentism and reductionism. In *Ecology revisited: Reflecting on concepts, advancing science*, eds. A. Schwarz and K. Jax, 31–43. Dordrecht: Springer.

Boulding, K.E. 1966. The economics of the coming spaceship earth. In *Environmental quality. In a growing economy: Essays from the sixth RFF forum*, ed. H. Jarrett, 3–14. Baltimore: The John Hopkins Press.

Brandon, R.N. 1988. The levels of selection: A hierarchy of interactors. In *The role of behavior in evolution*, ed. H.C. Plotkin, 51–71. Cambridge, MA: MIT Press.

Brandon, R.N. 1990. *Adaptation and environment*. Princeton: Princeton University Press.

Brandon, R.N. 1992. Environment. In *Keywords in evolutionary biology*, eds. E.F. Keller and E.A. Lloyd, 81–86. Cambridge, MA: Harvard University Press.

Callicott, J.B. 1986. On the intrinsic value of non-human species. In *The preservation of species. The value of biological diversity*, ed. B.G. Norton, 138–172. Princeton: Princeton University Press.

Callicott, J.B. 1989. *In defense of the land ethic: Essays in environmental philosophy*. Albany: State University of New York Press.
Carter, A. 2001. Review of Robin Attfield, The Ethics of the global environment. *Mind* 110: 149–153.
Clements, F.E. 1905/1977. *Research methods in ecology*. Lincoln: The University Publishing Co. Repr., New York: Arno Press.
Clements, F.E. 1907. *Plant physiology and ecology*. New York: Henry Holt and Company.
Clements, F.E. 1916. *Plant succession: An analysis of the development of vegetation*. Washington, DC: Carnegie Institution of Washington.
Clements, F.E. 1935. Experimental ecology in the public service. *Ecology* 16(3): 342–363.
Clements, F.E., and V.E. Shelford. 1939. *Bio-ecology*. New York: Wiley.
Cournot, A.-A. 1843/1984. *Exposition de la théorie des chances et des probabilités*. Paris: Vrin. 1st ed., published by Librairie de L. Hachette, Paris.
D'Ancona, U. 1939/1954. *The struggle for existence*. Trans. A. Charles and R.F.J. Withers. Leiden: E.J. Brill. 1st ed., published by Borntraeger, Berlin.
Dawkins, R. 1976/1989. *The selfish gene*. Oxford/New York: Oxford University Press.
Dewey, J. 1898/1972. Evolution and ethics. *The Monist* 8(3): 321–341. Repr., *The early works of John Dewey 1882–1898*, vol. 5, 1895–1898, ed. J.A. Boydston, 34–53. Carbondale and Edwardsville: Southern Illinois University Press.
Dewey, J. 1929/1984. *The quest for certainty: A study of the relation of knowledge and action*. New York: Minton & Balch Co. Repr., *The later works of John Dewey 1925–1953*, vol. 4, 1929, ed. J.A. Boydston, 1–250. Carbondale and Edwardsville: Southern Illinois University Press.
Dewey, J. 1930/1984. Conduct and experience. In *Psychologies of 1930*, ed. C. Murchison, 409–422. Worcester: Clark University Press. Repr., *The later works of John Dewey 1925–1953*, vol. 5, 1929–1930, ed. J.A. Boydston, 218–235. Carbondale and Edwardsville: Southern Illinois University Press.
Dewey, J. 1944/1989. Some questions about value. *The Journal of Philosophy* 41(17): 449–455. Repr., *The later works of John Dewey 1925–1953*, vol. 15, 1942–1948, ed. J.A. Boydston, 101–108. Carbondale and Edwardsville: Southern Illinois University Press.
Dewey, J., and A.F. Bentley. 1949/1989. *Knowing and the known*. Boston: The Beacon Press. Repr., *The later works of John Dewey 1925–1953*, vol. 16, 1949–1952, ed. J.A. Boydston, 1–294. Carbondale and Edwardsville: Southern Illinois University Press.
Eble, G.J. 1999. On the dual nature of chance in evolutionary biology and paleobiology. *Paleobiology* 25(1): 75–87.
El-Hani, C.N. 2007. Between the cross and the sword: The crisis of the gene concept. *Genetics and Molecular Biology* 30: 297–307.
Elton, C. 1927. *Animal ecology*. With an introduction by J.S. Huxley. New York: Macmillan.
Emlen, J.M. 1973. *Ecology: An evolutionary approach*. Reading: Addison-Wesley.
Fisher, R.A. 1924. The biometrical study of heredity. *The Eugenics Review* 16: 189–210.
Fisher, R.A. 1930. *The genetical theory of natural selection*. Oxford: Clarendon.
Gault, R. 1995. In and out of time. *Environmental Values* 4: 149–166.
Gause, G.F. 1934. *The struggle for existence*. Baltimore: Williams & Wilkins.
Geisler, N.L. 1971. *Ethics: Alternatives and issues*. Grand Rapids: Zondervan.
Gleason, H.A. 1917. The structure and development of the plant association. *Bulletin of the Torrey Botanical Club* 44(10): 463–481.
Gleason, H.A. 1926. The individualistic concept of the plant association. *Bulletin of the Torrey Botanical Club* 53(1): 7–26.
Godfrey-Smith, P. 2000. The replicator in retrospect. *Biology and Philosophy* 15: 403–423.
Gould, S.J. 1999. Introduction: The scales of contingency and puntuation in history. In *Structure and contingency: Evolutionary processes in life and human society*, ed. J. Bintliff, ix–xxii. London: Leicester University Press.

Griffiths, P.E., and R.D. Gray. 1997. Replicator II – Judgement day. *Biology and Philosophy* 12: 471–492.

Grinnell, J. 1917. The niche-relationships of the California Thrasher. *The Auk* 34(4): 427–433.

Hagen, J.B. 1992. *An entangled bank*. New Brunswick: Rutgers University Press.

Hamilton, W.D. 1964. The genetical evolution of social behavior II. *Journal of Theoretical Biology* 7: 17–52.

Hull, D.L. 1979. Philosophy of biology. In *Current research in philosophy of science*, eds. P.D. Asquith and H.E. Kyburg, 421–435. East Lansing: Philosophy of Science Association.

Hull, D.L. 1980. Individuality and selection. *Annual Review of Ecology and Systematics* 11: 311–332.

Hull, D.L. 1981. Units of evolution: A metaphysical essay. In *The philosophy of evolution*, eds. U.J. Jensen and R. Harré, 23–44. Brighton: The Harvester Press.

Hull, D.L. 1988/2001. Interactors versus vehicles. In *The role of behavior in evolution*, ed. H.C. Plotkin, 19–50. Cambridge, MA: MIT Press. Repr., *Science and selection. Essays on biological evolution and the philosophy of science*, ed. D.L. Hull, 13–45. Cambridge, UK: Cambridge University Press.

Hull, D.L. 1998. Introduction to part III. In *The philosophy of biology*, eds. D.L. Hull and M. Ruse, 149–152. Oxford/New York: Oxford University Press.

Hume, D. 1748/2007. *An enquiry concerning human understanding*. Edited with an introduction and notes by P. Millican. Oxford/New York: Oxford University Press. 1st ed., published as *Philosophical essays concerning human understanding*. London: A. Millar.

Hutchinson, G.E. 1959. Homage to Santa Rosalia or why are there so many kinds of animals? *The American Naturalist* 93: 145–159.

Hutchinson, G.E. 1965. *The ecological theater and the evolutionary play*. New Haven: Yale University Press.

Huxley, T.H. 1894/1920. Evolution and ethics, and other essays. In *Collected essays*, 9 vols., vol. 9. London/New York: Macmillan. Repr., New York/London: D. Appleton and Co.

Huxley, T.H. 1907. *Aphorisms and reflections*. From the works of T.H. Huxley, selected by H.A. Huxley. London: MacMillan.

Inchausti, P. 1994. Reductionist approaches in community ecology. *The American Naturalist* 143(2): 201–221.

Kant, I. 1775–1780/1963. *Lectures on ethics*. Trans. L. Infield and foreword by L.W. Beck. Indianapolis: Hackett Publishing Company.

Kant, I. 1785/1990. *Foundations of the metaphysics of morals and, What is enlightenment?* Trans. with an introduction by L.W. Beck. New York/London: Macmillan/Collier Macmillan. Original editions: *Grundlegung zur Metaphysik der Sitten* and *Beantwortung der Frage: Was ist Aufklärung?*

Katz, J. 1944. On chance and prediction. *The Journal of Philosophy* 41(23): 626–631.

Kingsland, S.E. 1985/1995. *Modeling nature: Episodes in the history of population ecology*. Chicago: University of Chicago Press.

Kropotkin, P.A. 1902/1972. *Mutual aid. A factor of evolution*. Edited and with an introduction by P. Avrich. New York: New York University Press. 1st ed., published by Heinemann, London.

Leopold, A. 1949/1966. *A Sand County almanac. With other essays on conservation from Round River*. New York: Oxford University Press.

Lewontin, R.C. 1970. The units of selection. *Annual Review of Ecology and Systematics* 1: 1–18.

Lewontin, R.C. 1991. *Biology as ideology*. New York: Harper.

Lindeman, R.L. 1941. Seasonal food-cycle dynamics in a senescent lake. *American Midland Naturalist* 26(3): 636–673.

Lindeman, R.L. 1942. The trophic-dynamic aspect of ecology. *Ecology* 23: 399–417.

Lotka, A.J. 1925/1956. *Elements of physical biology*. Baltimore: Williams & Wilkins. Repr., *Elements of mathematical biology*. New York: Dover.

Maynard-Smith, J. 1964. Group selection and kin selection. *Nature* 201: 1145–1146.

Maynard-Smith, J. 1976. Group selection. *The Quarterly Review of Biology* 51: 277–283.

Maynard-Smith, J. 1998. Review of Elliott Sober and David Sloan Wilson, Unto others: The evolution and psychology of unselfish behavior. *Nature* 393: 639–640.

McCloskey, H.J. 1965. Rights. *The Philosophical Quarterly* 15(59): 115–127.
McIntosh, R.P. 1980/1982. The background and some current problems of theoretical ecology. *Synthese* 43(2): 195–255. Repr., *Conceptual issues in ecology*, ed. E. Saarinen, 1–62. Dordrecht: D. Reidel.
McIntosh, R.P. 1985. *The background of ecology. Concept and theory*. Cambridge, UK: Cambridge University Press.
McIntosh, R.P. 1987. Pluralism in ecology. *Annual Review of Ecology and Systematics* 18: 321–341.
Meyer, L.M.N., G.C. Bomfim, and C.N. El-Hani. 2011. How to understand the gene in the twenty-first century? *Science & Education* 2011. Doi:10.1007/s11191-011-9390-z.
Mill, J.S. 1848/1929. *Principles of political economy with some of their applications to social philosophy*. Edition with an introduction by W.J. Ashley. London/New York: Longmans, Green. 1st ed., published by J.W. Parker and Son, West Strand, London.
Moore, G.E. 1903/2000. *Principia Ethica*. With an introduction by ed. T. Baldwin, revised ed. with preface to the 2nd ed. and other papers. Cambridge, UK/New York: Cambridge University Press.
Moore, G.E. 1922. *Philosophical Studies*. London: Routledge and Kegan Paul.
Morange, M. 2001. *The misunderstood gene*. Cambridge, MA: Harvard University Press.
Morgan, T.H. 1910. Chance or purpose in the origin and evolution of adaptation. *Science* 31(new series 789): 201–210.
Norton, B.G. 1987. *Why preserve natural variety?* Princeton: Princeton University Press.
Norton, B.G. 2003. *Searching for sustainability: Interdisciplinary essays in the philosophy of conservation biology*. Cambridge, UK/New York: Cambridge University Press.
Norton, B.G. 2005. *Sustainability: A philosophy of adaptive ecosystem management*. Chicago: University of Chicago Press.
Norton, B.G. 2007. The past and future of environmental ethics/philosophy. *Ethics & the Environment* 12(2): 134–136.
O'Neill, J. 1992. The varieties of intrinsic value. *The Monist* 75(2): 119–137.
Okasha, S. 2006. *Evolution and the levels of selection*. Oxford, UK/New York: Oxford University Press.
Oyama, S. 1985/2000. *The ontogeny of information: Developmental systems and evolution*. Cambridge, UK: Cambridge University Press. 2nd ed. revised and expanded with a foreword by R.C. Lewontin. Durham: Duke University Press.
Palka-Santini, M., and L. Palka. 1997. Microbial ecosystem in human or animal? *Bulletin of the Ecological Society of America* 78: 298–299.
Passmore, J. 1974. *Man's responsibility for nature: Ecological problems and western traditions*. London: Duckworth.
Phillips, J. 1931. The biotic community. *Journal of Ecology* 19(1): 1–24.
Pianka, E.R. 1974/2000. *Evolutionary ecology*. New York: Harper and Row. 6th ed. San Francisco: Benjamin-Cummings/Addison-Wesley-Longman.
Pianka, E.R. 2005. Species diversity. In *Encyclopedia of evolution*, ed. M. Pagel, e-reference ed. Oxford, UK: Oxford University Press. Florida State University. 6 July 2011 http://www.oxford-evolution.com/entry?entry=t169.e396.
Price, M.V. 1986. Introduction to the symposium: Mechanistic approaches to the study of natural communities. *American Zoologist* 26(1): 3–4.
Ratner, S., and J. Altman (eds.). 1964. *John Dewey and Arthur F. Bentley. A philosophical correspondence, 1932–1951*. New Brunswick: Rutgers University Press.
Regan, T. 1983. *The case for animal rights*. Berkeley/Los Angeles: University of California Press.
Regan, T. 1985. The case for animal rights. In *In defense of animals*, ed. P. Singer, 13–26. New York: Basil Blackwell.
Rolston, H. III. 1985. Duties to endangered species. *BioScience* 35(11): 718–726.
Rolston, H. III. 1988. *Environmental ethics: Duties to and values in the natural world*. Philadelphia: Temple University Press.
Roughgarden, J., and J.M. Diamond. 1986. Overview: The role of species interactions in community ecology. In *Community ecology*, eds. J.M. Diamond and T.J. Case, 333–343. New York: Harper & Row.

Ruse, M. 1995. Evolutionary ethics. A defense. In *Biology, ethics, and the origins of life*, ed. H. Rolston III, 89–112. Boston: Jones and Bartlett Publishers.
Ruse, M. 2008. Chance and evolution. In *Creation: Law and probability*, ed. F. Watts, 110–123. Aldershot: Ashgate Publishing Company.
Ruse, M. (ed.). 2009. Introduction. In *Thomas Henry Huxley: Evolution & ethics*. Edited and with an introduction by M. Ruse, vii–xxxvi. Princeton/Oxford: Princeton University Press.
Salt, H.S. 1894. *Animals' rights considered in relation to social progress*. London/New York: Macmillan.
Schoener, T.W. 1986. Mechanistic approaches to community ecology: A new reductionism? *American Zoologist* 26(1): 81–106.
Simberloff, D. 1980/1982. A succession of paradigms in ecology: Essentialism to materialism and probabilism. *Synthese* 43(1): 3–39. Repr., *Conceptual issues in ecology*, ed. E. Saarinen, 63–100. Dordrecht: D. Reidel.
Simpson, G.G. 1949. *The meaning of evolution. A study of the history of life and of its significance for man*. New Haven: Yale University Press.
Singer, P. 1977. *Animal liberation*. New York: Avon Books.
Sober, E. 1984. *The nature of selection: Evolutionary theory in philosophical focus*. Cambridge, MA: MIT Press.
Sober, E., and D.S. Wilson. 1998. *Unto others: The evolution and psychology of unselfish behavior*. Cambridge, MA: Harvard University Press.
Spencer, H. 1892–1893. *The principles of ethics*, 2 vols. New York: Appleton and Company.
Stone, C.D. 1972. Should trees have standing? Toward legal rights for natural objects. *Southern California Law Review* 45: 450–501.
Strong, D.R. Jr., D. Simberloff, L.G. Abele, and A.B. Thistle (eds.). 1984. *Ecological communities. Conceptual issues and the evidence*. Princeton: Princeton University Press.
Tansley, A.G. 1935. The use and abuse of vegetational concepts and terms. *Ecology* 16: 284–307.
Taylor, P.W. 1986. *Respect for nature: A theory of environmental ethics*. Princeton: Princeton University Press.
Taylor, W.P. 1936. What is ecology and what good is it? *Ecology* 17(3): 333–346.
Volterra, V. 1926. Fluctuations in the abundance of a species considered mathematically. *Nature* 118: 558–560.
Volterra, V., and U. D'Ancona. 1935. *Les associations biologiques au point de vue mathématique*. Paris: Hermann.
Warren, H.C. 1916. Purpose, chance, and other perplexing concepts. *The Journal of Philosophy* 13(16): 441–442.
Whittaker, R.H. 1956. Vegetation of the Great Smoky Mountains. *Ecological Monographs* 26(1): 1–80.
Whittaker, R.H., and S.A. Levin. 1975. *Niche: Theory and application*. Stroudsburg: Dowden, Hutchinson & Ross.
Williams, G.C. 1966. *Adaptation and natural selection: A critique of some current evolutionary thought*. Princeton: Princeton University Press.
Wilson, D.S. 1975. A theory of group selection. *Proceedings of the National Academy of Sciences USA* 72: 143–146.
Wilson, D.S. 1983. The group selection controversy: History and current status. *Annual Review of Ecology and Systematics* 14: 159–187.
Wilson, D.S. 2001. Cooperation and altruism. In *Evolutionary ecology: Concepts and case studies*, eds. C.W. Fox, A.R. Derek, and J.F. Daphne, 222–231. Oxford/New York: Oxford University Press.
Wynne-Edwards, V.C. 1962. *Animal dispersion in relation to social behaviour*. New York: Hafner Publishing Company.

Chapter 2
Evolution Versus Creation: A Sibling Rivalry?

Michael Ruse

Abstract In this paper, I argue that evolutionary thinking started as a secular response to the Christianity of the eighteenth century. While I agree that Charles Darwin's theory of evolution was in essence scientific, I argue that Darwin's supporters often wanted to continue to treat evolutionary thinking as a secular response to religious claims. This continues to the present, and I suggest that evolutionists should be aware of this fact and temper their thinking and behavior accordingly.

Q: Dr. Ruse, having examined the creationist literature at great length, do you have a professional opinion about whether creation science measures up to the standards and characteristics of science that you have just been describing?

A: Yes, I do. In my opinion, creation science does not have those attributes that distinguish science from other endeavours.

Q: Would you please explain why you think it does not.

A: Most importantly, creation science necessarily looks to the supernatural acts of a Creator. According to creation-science theory, the Creator has intervened in supernatural ways using supernatural forces.

Q: Do you think that creation science is testable?

A: Creation science is neither testable nor tentative. Indeed, an attribute of creation science that distinguishes it quite clearly from science is that it is absolutely certain about all of the answers. And considering the magnitude of the questions it addresses – the origins of man, life, the earth, and the universe – that certainty is all the more revealing. Whatever the contrary evidence, creation science never accepts that its theory is falsified. This is just the opposite of tentativeness and makes a mockery of testing.

Q: Do you find that creation science measures up to the methodological considerations of science?

A: Creation science is woefully lacking in this regard. Most regrettably, I have found innumerable instances of outright dishonesty, deception, and distortion used to advance creation-science arguments.

M. Ruse (✉)
Department of Philosophy, Florida State University, Tallahassee, FL 32306, USA
e-mail: mruse@fsu.edu

D. Bergandi (ed.), *The Structural Links between Ecology, Evolution and Ethics: The Virtuous Epistemic Circle*, Boston Studies in the Philosophy of Science 296,
DOI 10.1007/978-94-007-5067-8_2, © Springer Science+Business Media Dordrecht 2013

Q: Dr. Ruse, do you have an opinion to a reasonable degree of professional certainty about whether creation science is science?
A: Yes.
Q: What is your opinion?
A: In my opinion creation science is not science.
Q: What do you think it is?
A: As someone also trained in the philosophy of religion, in my opinion creation science is religion (Ruse 1988, 304–306).

My moment of triumph! The time was December 1981 and the place was Little Rock, Arkansas. The occasion was a court trial, brought on by the American Civil Liberties Union, that organization dedicated to the defense of the US Constitution. It was attacking a new law in the state that insisted that children be taught, alongside evolution, something called (by its defenders) Creation Science and better known to the rest of us as Genesis taken absolutely literally. I was an expert witness for the ACLU, and thanks to my testimony and that of others (notably including the late Stephen Jay Gould), we won a terrific victory. Creation Science was ruled to be religion and hence not admissible to publicly funded schools, and that was the end of that. For once in his life, a philosopher had shown that he was not entirely useless.

That was 30 years ago, and time has shown that we evolutionists celebrated a little too quickly. Crude Biblical literalism may have been vanquished, but an evangelical-Christian-inspired approach to nature is still alive and well – these days often under the label of Intelligent Design – and evolutionists continue to have to fight for the theory that they love so deeply. Anyone who says confidently that the schoolchildren of the United States will never learn about Noah's Flood in biology classes in the twenty-first century sees ahead more clearly and confidently than I. The usual feeling of evolutionists – certainly the feeling that I had for many years before, during, and after Arkansas – is that it is a simple matter of right and wrong, black and white. The Christians are wrong and the evolutionists are right. The world was not made in 6 days, 6,000 years ago. Adam and Eve were not made miraculously. There was no universal flood. Rather, everything living is the end result of a long, slow, natural process of development, and (although there is some debate about its extent) the chief cause is that identified by Charles Darwin in his *Origin of Species*: natural selection brought on by a struggle for existence.

I still think that this is basically the truth, and please note that nothing I am about to say in any way should be taken as a weakening of my convictions in this respect. I am with Darwin all of the way. I agree with the philosopher Dan Dennett (1995), who has said that natural selection is the great idea of all time. But I do think now that there is more to the story than good and bad, Heaven and Hell. I believe – and in this paper I shall try to justify this belief – that there is a good reason why, over and above the simple facts of the case, the evangelicals are tense about Darwinism (using the term generically for evolutionism). As an evolutionist, I look to the past for solutions and understanding, and I believe that by looking back at the history of evolution we see that it has always been more than just science – and continues to be more to this day – and that this excess is of a kind directly to challenge those with religious convictions. In short, I argue that evolution has itself (and still does) function as a challenge to conventional religion – it is, if you will, a secular religion – and there is no wonder that the Creationists and fellow travelers get so het up. I am not the first to say this. The Creationists have been saying it for some time. But I think I am the first – or one of the first – to say it from the evolutionary side. I am not a traitor – at least, I do not think of myself in this way – even though I am saying naught for our comfort. Without wanting to sound like a pretentious prig, my hope indeed is to show to us evolutionists what we are doing. I do not expect or really want people to change their minds about their beliefs, but I do want to show when it is appropriate to make claims of one kind and when it is appropriate to make claims of another kind. And when perhaps we should keep our mouths shut, although there are no doubt those who think that I should be the first to take my own advice.

2.1 Before The Origin

Evolution is an idea with its roots in the eighteenth century. This was the time when the ideology of progress – the belief that humans through their own unaided efforts can change and improve their lots – became dominant, and there were several who took the cultural idea of progress and read it into the rocks, thereby making for an evolutionary or transmutationary view of life's history. Usually, they then promptly took their evolutionism and argued in a good circular fashion that this justified their beliefs in progress (Darwin 1803)! This continued right through to the middle of the nineteenth century. The poet Alfred Tennyson shows this in his famous and much-loved poem *In Memoriam*. This poem was begun in the 1830s but not completed until about 1850. It is a testament to the memory of a young friend of Tennyson, Arthur Henry Hallam, whose life was cut short. Tennyson writes at first in the poem about his desolation at Hallam's death and existence's apparent meaninglessness, something which he found reflected in the uniformitarian geology of Charles Lyell. Lyell had argued that nature is going nowhere, just simply bound by unbroken stern laws, and that there is no end in prospect, nor any progress in view. Life comes and life goes without meaning as expressed in the following famous passage:

> Are God and Nature then at strife,
>   That Nature lends such evil dreams?
>   So careful of the type she seems,
> So careless of the single life;
>
>   ....
>
> So careful of the type? but no.
>   From scarped cliff and quarried stone
>   She cries, "A thousand types are gone:
> I care for nothing, all shall go. "

Given Nature "red in tooth and claw" – this is the source of this famous phrase – nothing seems to make any sense. Not only are individuals pointless mortals, but so also are groups. We are born, we live, and then we die – usually painfully. Nothing makes sense or has meaning. There are just endless Lyellian cycles. Then towards the end of the 1840s Tennyson read Chambers, or at least he read a very detailed review of Chambers's *Vestiges*. Chambers (1844) argued for an organic evolution that was unambiguously progressionist, that is to say, moving up from simple forms to humans, and then perhaps beyond. Inspired by this, Tennyson picked up pen and finished his poem. He argued in the final lines that perhaps there is meaning after all, despite a Lyellian uniformitarianism: that life is progressing upwards, and that perhaps will go on beyond the human form that we have at present. Could it not be that Hallam represented some anticipation of the more-developed life to come, cut short as it were in its prime? There is therefore hope for us all and a meaning for the life of Hallam.

> A soul shall strike from out the vast
> And strike his being into bounds,
> And moved thro' life of lower phase,
> Result in man, be born and think,
> And act and love, a closer link
> Betwixt us and the crowning race
> ....
> Whereof the man, that with me trod
> This planet, was a noble type
> Appearing ere the times were ripe,
> That friend of mine who lives in God.

2.2 Charles Darwin

So much for evolution before 1859, the year in which *On the Origin of Species* was published. What did Darwin do and how did he alter things? Start with what he did. He set out to give a new theory of evolution, one that could indeed stand muster against a proper empirical approach to science. He made the fact of evolution secure, and he proposed the mechanism – natural selection – that is generally considered by scientists today to be the key factor behind the development of organisms, i.e., a development by a slow natural process from a few simple forms, and perhaps indeed ultimately from inorganic substances (although, sagely, Darwin said nothing on this latter topic). In *The Origin*, after first stressing the analogy between the world of the breeder and the world of nature, and after showing how much variation exists between organisms in the wild, Darwin was then ready for the key inferences. First, an argument to the struggle for existence and, following on this, an argument to the mechanism of natural selection.

> Let it be borne in mind in what an endless number of strange peculiarities our domestic productions, and, in a lesser degree, those under nature, vary; and how strong the hereditary tendency is. Under domestication, it may be truly said that the whole organization becomes in some degree plastic. Let it be borne in mind how infinitely complex and close-fitting are the mutual relations of all organic beings to each other and to their physical conditions of life. Can it, then, be thought improbable, seeing that variations useful to man have

> undoubtedly occurred, that other variations useful in some way to each being in the great and complex battle of life, should sometimes occur in the course of thousands of generations? If such do occur, can we doubt (remembering that many more individuals are born than can possibly survive) that individuals having any advantage, however slight, over others, would have the best chance of surviving and of procreating their kind? On the other hand we may feel sure that any variation in the least degree injurious would be rigidly destroyed. This preservation of favourable variations and the rejection of injurious variations, I call Natural Selection (Darwin 1859, 80–81).

With the main mechanisms of change thus presented, Darwin introduced the famous metaphor of a tree. "The affinities of all the beings of the same class have sometimes been represented by a great tree. I believe this simile largely speaks the truth." The leaves and twigs at the top represent the species extant today. Then as we go down the branches, we have the great evolutionary paths of yesterday. All the way down we go until we reach the very first shared origins of life. "As buds give rise by growth to fresh buds, and these, if vigorous, branch out and overtop on all sides many a feebler branch, so by generation I believe it has been with the great Tree of Life, which fills with its dead and broken branches the crust of the earth, and covers the surface with its ever branching and beautiful ramifications" (Darwin 1859, 129–130).

Then from this, Darwin turned to a general survey of the biological world, offering what the philosopher William Whewell (1840) had dubbed a "consilience of inductions." Each area was explained by evolution through natural selection, and in turn each area contributed to the support of the mechanism of evolution through natural selection. Geographical distribution (biogeography) was a triumph, as Darwin explained just why it is that one finds the various patterns of animal and plant life around the globe. Why, for instance, does one have the strange sorts of distributions and patterns that are exhibited by the Galapagos Archipelago and other island groups? It is simply that the founders of these isolated island denizens came by chance from the mainlands, and once established started to evolve and diversify under the new selective pressures to which they were now subject. Embryology likewise was a particular point of pride for Darwin. Why is it that the embryos of some different species are very similar – man and the dog for instance – whereas the adults are very different? Darwin argued that this follows from the fact that in the womb the selective forces on the two embryos would be very similar – they would not therefore be torn apart – whereas the selective forces on the two adults would be very different – they would be torn apart. Here, as always through his discussions of evolution, Darwin turned to the analogy with the world of the breeders in order to clarify and support the point at hand. "Fanciers select their horses, dogs, and pigeons, for breeding, when they are nearly grown up: they are indifferent whether the desired qualities and structures have been acquired earlier or later in life, if the full-grown animal possesses them" (Darwin 1859, 446). And finally, all of this led to that famous passage at the end of *The Origin*: "There is a grandeur in this view of life, with its several powers, having been originally breathed into a few forms or into one; and that, whilst this planet has gone cycling on according to the fixed law of gravity, from so simple a beginning endless forms most beautiful and most wonderful have been, and are being, evolved" (Darwin 1859, 490).

So much for the theory. Now, in the light of the history presented thus far, what was Darwin hoping to do? Two things we can say immediately. He was not repudiating progress. It may have had a somewhat subdued role, but as the quotation just given at the end of the last paragraph shows unambiguously, biological progress was there and believed in. So I am certainly not saying that Darwin broke absolutely with his past. Indeed, in a way I am hinting that if someone were a Christian perhaps, for the first time, here was an evolutionary theory that might be molded and adapted for use without giving up one's faith. But, in the context of this present discussion, I think more important than the continuities was Darwin's determination to make of his theory something with a different status from those of his predecessors. Darwin did not want to produce a secular religion. He wanted to produce a functioning, empirical science. He wanted something, to use the language of Thomas Kuhn (1962), that could work as a "paradigm," making possible normal science. The kind of normal science that in fact he himself was to do soon after *The Origin*, when he wrote a little book on orchids (Darwin 1862) and that others were to do, like Henry Walter Bates (1862, 511–513) when he used natural selection to produce an explanation of butterfly mimicry. Progress was there, but it was downplayed. References may have been made to the Creator, but He was given no work to do, and could have been dropped without loss of content. Evolution through natural selection was certainly going to contradict Genesis taken literally, but to think that Darwin was offering a "religion without revelation" (to borrow a title from a book of the twentieth century) would be quite to misunderstand his intent (see Huxley 1927).

2.3 The Darwinian Evangelist

So, what happened? I argue that Darwinism got hijacked, and turned to other purposes (Ruse 1996, 1999a, 1999b, 2000). And the chief hijacker was none other than he who is celebrated as "Darwin's Bulldog," the nineteenth-century morphologist and paleontologist Thomas Henry Huxley. Unlike Darwin – a rich man, sick for most of his adult life, able to live as a semi-recluse – Huxley was a man who was making his own way, as a university professor and then as a college dean. He, with a number of others (mainly men but with some women like Florence Nightingale), were striving hard to change the course of British life, away from the near-feudalism of the rural eighteenth century and towards the modern, urban industrialism of the twentieth century. They were reforming the civil service, the military, the medical profession, and more – including teaching at school and university. Huxley was in the thick of creating a professional science – a professional science where one could succeed on merit and make a living – and Huxley realized full well that to achieve his aims he had to find reasons to employ the young scientists he was producing. Physiology he sold to the medical profession, arguing (with success) that the time had come to stop killing people and to start curing. Morphology he sold to the teaching profession, which was at a crucial point for only now was education starting to become the birthright of all and not under the sway of organized religion. For evolution, alas, Huxley could see no immediate cash value. It cured no pains in the belly, and it was too daring for the junior classroom. But Huxley – a dedicated evolutionist, albeit somewhat indifferent to natural selection – could nevertheless see a role for evolution. It would be the ideology – the secular religion – of the reformers, being something to put against the ideology – the spiritual religion – allied with those who resisted change. It would be the system giving answers to origins and explaining the status of humankind to replace the outdated system of the conservatives and reactionaries, who worshiped each Sunday in the local Anglican parish church. Evolution versus Christianity.

Progress, naturally, was to be the backbone of the system. But more was needed. A good religion has a moral system, a set of ethical prescriptions – Thou shalt not kill; Love your neighbour as yourself; and so forth. Charles Darwin was not really into this sort of thing, but there was another English evolutionist ready and very willing to step into the breach. Herbert Spencer's evolutionism starts (continues and finishes) with progress. For him, progress was not so much an empirical finding but a metaphysical presupposition of his view of history. It ran through everything, from the most primitive forms of culture to the evolution of our own species.

> Now, we propose in the first place to show, that this law of organic progress is the law of all progress. Whether it be in the development of the Earth, in the development of Life upon its surface, in the development of Society, of Government, of Manufactures, of Commerce, of Language, Literature, Science, Art, this same evolution of the simple into the complex, through a process of continuous differentiation, holds throughout. From the earliest traceable cosmical changes down to the latest results of civilization, we shall find that the transformation of the homogeneous into the heterogeneous, is that in which Progress essentially consists (Spencer 1857, 446–447).

Morality fit nicely into all of this. It is our obligation to preserve and to promote progress. Here there is a place for the struggle and selection. Even in 1851 some years before *The Origin* was published, Spencer speculated on selective effects showing themselves in the different natures and behaviours of the Irish and the Scots. He concluded that struggle and selection in society translates into extreme laissez-faire socioeconomics: the state should stay out of the way of people pursuing their own self-interests and should not at all attempt to regulate practices or redress imbalances or unfairnesses. Libertarian license therefore is not only the way that things are, but the way that they should be. In fact, Spencer was far from convinced that mid-Victorian Britain was a laissez-faire society, but this is what he hoped fervently that it would become.

> ... we must call those spurious philanthropists, who, to prevent present misery, would entail greater misery upon future generations. All defenders of a poor law must, however, be classed amongst such. That rigorous necessity which, when allowed to act on them, becomes so sharp a spur to the lazy, and so strong a bridle to the random, these pauper's friends would repeal, because of the wailings it here and there produces. Blind to the fact, that under the natural order of things society is constantly excreting its unhealthy, imbecile, slow, vacillating, faithless members, these unthinking, though well-meaning, men advocate

> an interference which not only stops the purifying process, but even increases the vitiation — absolutely encourages the multiplication of the reckless and incompetent by offering them an unfailing provision, and *dis*courages the multiplication of the competent and provident by heightening the prospective difficulty of maintaining a family (Spencer 1851, 323–324).

The point I make, therefore, is that Charles Darwin was both a success and a failure. He was a success inasmuch (and it is a very big "inasmuch") as he turned people to evolution. Before him, it had been a pseudo-scientific idea, on a par with astrology or phrenology. (Interestingly, Chambers had started to write a book on phrenology – the science of brain bumps – and changed half-way through to writing a book on evolution.) After Darwin, evolution was common sense. He was a failure inasmuch (and you judge how big an "inasmuch" you think this to be) he did not turn evolution into a functioning, professional science, with natural selection at its heart. Evolution was a raging success, but more in a bastardized Spencerian version, functioning less as a science and more as a secular religion (Bannister 1979). That was what the reformers like Huxley wanted and that was what the reformers like Huxley got. When Jesus died on the cross, there was no religion of Christianity. That was for St Paul to create, and people have been arguing ever since about the relationship between the life and teachings of Jesus and the religion that St Paul left behind. When Darwin wrote *The Origin*, there was no science of Darwinism. That was for Thomas Henry Huxley to create, and I argue that the relationship between the teachings of Darwin and the religion of Huxley was about as iffy as that between Jesus and Paul.

2.4 The Twenty-first Century

I will skip much subsequent history, coming down to the present, summing up, and drawing a conclusion. I argue strongly and strenuously that there is today a mature evolutionary biology – Darwin-based, empirical, predictive, explanatory. It has felt and benefitted from the full blast of the molecular revolution in biology, and it looks forward into this new century with great accomplishments, with powerful tools, and with an anticipation of solving major problems old and new. I mention simply as illustration the incredible advances over the past two decades in the understanding of development and of how this is now being integrated into the evolutionary picture (so-called "evo-devo"). This evolutionary biology is not, by any stretch of the imagination, a secular religion, and those who quote me as saying that it is (or pretend that I have not mentioned and stressed its existence and importance) do me and evolutionary biology a grave disfavour. But, given our history, you would expect more to the story, and indeed there is. I would argue also that – in the tradition of Thomas Henry Huxley – there is another side that continues unabated today. And this side does use evolution as a secular religion.

And do not underestimate this side. Many who play this game are great evolutionary biologists in their own right. One thinks here of the distinguished Harvard entomologist and sociobiologist Edward Osborne Wilson, who has made major advances in our understanding of social behaviour. He nevertheless is explicit in wanting to make more of his science than mere science. Consider for instance the use he makes of evolution in his Pulitzer Prize-winning *On Human Nature*:

> But make no mistake about the power of scientific materialism. It presents the human mind with an alternative mythology that until now has always, point for point in zones of conflict, defeated traditional religion. Its narrative form is the epic: the evolution of the universe from the big bang of fifteen billion years ago through the origin of the elements and celestial bodies to the beginnings of life on earth. The evolutionary epic is mythology in the sense that the laws it adduces here and now are believed but can never be definitively proved to form a cause-and-effect continuum from physics to the social sciences, from this world to all other worlds in the visible universe, and backward through time to the beginning of the universe. Every part of existence is considered to be obedient to physical laws requiring no external control. The scientist's devotion to parsimony in explanation excludes the divine spirit and other extraneous agents. Most importantly, we have come to the crucial stage in the history of biology when religion itself is subject to the explanations of the natural sciences. As I have tried to show, sociobiology can account for the very origin of mythology by the principle of natural selection acting on the genetically evolving material structure of the human brain.
>
> If this interpretation is correct, the final decisive edge enjoyed by scientific naturalism will come from its capacity to explain traditional religion, its chief competitor, as a wholly material phenomenon. Theology is not likely to survive as an independent intellectual discipline (Wilson 1978, 192).

Like Spencer (a thinker whom Wilson admires greatly), over the years Wilson has offered all sorts of moral prescriptions, most particularly about the need to preserve biodiversity and to cherish the plants of the world, especially those vanishing from the Brazilian rainforests (where Wilson has spent much of his professional life). And it will not surprise the reader to find that progress is the force and reason behind everything: "the overall average across the history of life has moved from the simple and few to the more complex and numerous. During the past billion years, animals as a whole evolved upward in body size, feeding and defensive techniques, brain and behavioral complexity, social organization, and precision of environmental control – in each case farther from the nonliving state than their simpler antecedents did." (Wilson 1992, 187) Hence: "Progress, then, is a property of the evolution of life as a whole by almost any conceivable intuitive standard, including the acquisition of goals and intentions in the behavior of animals" (*Ibid.*). For Wilson, as for Spencer, progress confers value, and hence it is our obligation to promote (or at least not hinder) the evolutionary process.

What about my own position? It is not really relevant (although see Ruse 2001, 2003)! My intent here is simply to draw your attention to the fact that the tradition of making evolution into something more than a science – into a secular religion, to be blunt – is far from dead. It thrives on both sides of the Atlantic (and I am sure elsewhere, even – or perhaps especially – in post-Communist Russia). Anyone who thinks Wilson is not offering an evolution-based, Christianity alternative should read the pertinent passages again. And with my conclusion drawn, I will now allow myself a prescription. I do not want to stop Wilson and his fellows from doing what they do. Apart from anything else, as a historian and philosopher of science, I would be putting myself out of a job! And, too soon, people would be suggesting that I should not do what I do. But I do want to say to my fellow evolutionists, to my fellow Darwinians: Be aware of what you are doing, and do not pretend that you are doing straight science when you are not. Most particularly – and here I speak particularly to those of us who live and work in America – do not mix up your science and religion when you are teaching. It is illegal and unwise. The Creationists are out there watching what you are doing, and if you are teaching religion (secular or otherwise) under the guise of science, they will soon find out. And then they will have a lever, either to teach Creationism in the schools or to ban evolution from the schools. Either of these disjuncts would be a tragedy.

References

Bannister, R.C. 1979. *Social Darwinism: Science and myth in Anglo-American social thought.* Philadelphia: Temple University Press.
Bates, H.W. 1862. Contributions to an insect fauna of the Amazon valley. Lepidoptera: Heliconidæ. *Transactions of the Linnean Society of London* 23: 495–566.
Chambers, R. 1844. *Vestiges of the natural history of creation.* London: J. Churchill.
Darwin, C. 1859. *On the origin of species.* London: John Murray.
Darwin, C. 1862. *On the various contrivances by which British and foreign orchids are fertilised by insects, and on the good effects of intercrossing.* London: John Murray.
Darwin, E. 1803. *The temple of nature; or the origin of society: A poem, with philosophical notes.* London: J. Johnson, St. Paul's Church-Yard.
Dennett, D.C. 1995. *Darwin's dangerous idea.* New York: Simon & Schuster.
Huxley, J.S. 1927. *Religion without revelation.* London: Ernest Benn.
Kuhn, T. 1962. *The structure of scientific revolutions.* Chicago: University of Chicago Press.
Ruse, M. (ed.). 1988. *But is it science? The philosophical question in the creation/evolution controversy.* Buffalo: Prometheus.
Ruse, M. 1996. *Monad to man: The concept of progress in evolutionary biology.* Cambridge, MA: Harvard University Press.
Ruse, M. 1999a. *The Darwinian revolution: Science red in tooth and claw*, 2nd ed. Chicago: University of Chicago Press.
Ruse, M. 1999b. *Mystery of mysteries: Is evolution a social construction?* Cambridge, MA: Harvard University Press.
Ruse, M. 2000. *The evolution wars: A guide to the controversies.* Santa Barbara: ABC-CLIO.
Ruse, M. 2001. *Can a Darwinian be a Christian? The relationship between science and religion.* Cambridge: Cambridge University Press.
Ruse, M. 2003. *Darwin and design: Science, philosophy, religion.* Cambridge, MA: Harvard University Press.
Spencer, H. 1851. *Social statics: Or, the conditions essential to human happiness specified, and the first of them developed.* London: John Chapman (published in Dec 1950).
Spencer, H. 1857. Progress: Its law and cause. *Westminster Review* 67(new series 11(2)): 445–485.
Tennyson, A. 1850. *In memoriam.* London: Edward Moxon.
Whewell, W. 1840. *The philosophy of the inductive sciences, founded upon their history*, 2 vols. London: John W. Parker.
Wilson, E.O. 1978. *On human nature.* Cambridge, MA/London: Harvard University Press.
Wilson, E.O. 1992. *The diversity of life.* Cambridge, MA: Harvard University Press.

Chapter 3
Evolution and Chance

Jean Gayon

Abstract Chance comes into play at many levels in the explanation of the evolutionary process. This paper examines the various senses this concept takes at different levels, including mutation, genetic drift, genetic revolutions, ecosystems and macroevolution. Three main concepts of chance are identified: luck, randomness and contingency relative to a given theoretical system.

There is not, nor has there ever been, a biological theory that claims to explain the evolution of species by chance, without any further qualification. This idea only ever appears in a polemical context, where thinkers, philosophers or theologians who do not like a particular theory of evolution impute this notion to it. This does not mean that the notion of chance is not involved in the explanation of evolutionary phenomena. It has virtually always been present, notably in the successive versions of the Darwinian understanding of evolution.

The notion of chance is notoriously ambiguous, so much so that it makes little sense to speak in general about the role of chance in evolution. Two conditions must be fulfilled for the relation between chance and evolution to have meaning. We need to clarify what we mean by the word "chance" as well as the scientific contexts in which the various notions of chance are used. If these two conditions – analytical and contextual, or if you prefer, philosophical and scientific – are not met, all general declarations about the role of chance in biological evolution will be sterile.

I consider that contemporary evolutionary theory contains three possible meanings of the word "chance": luck, randomness and contingency with regard to a given theoretical system. These three terms – luck, randomness and contingency – are often confused with each other, and with "fortuitous." "Fortuitous" is the adjective that corresponds best to the substantive "chance." (In French, matters are confused by the

J. Gayon (✉)
Institute of History and Philosophy of Science and Techniques, University of Paris 1 (Panthéon Sorbonne), 75006, Paris, France
e-mail: jean.gayon@gmail.com

D. Bergandi (ed.), *The Structural Links between Ecology, Evolution and Ethics: The Virtuous Epistemic Circle*, Boston Studies in the Philosophy of Science 296,
DOI 10.1007/978-94-007-5067-8_3,

use of the term "*hasard*" for "chance" in an evolutionary context; the obvious, and ancient, English equivalent – "hazard" – has now changed its meaning.)

In the first part of my article, I will define the three concepts of "chance" mentioned above. In the second part, I will apply them to the main kind of phenomena that modern evolutionary theory characterizes and explains as being types of fortuitous events.

3.1 Three Meanings of the Concept of Chance

3.1.1 Luck

The most familiar meaning of the word "chance" appears in human contexts in which something happens that is unexpected, rather than the consequence of a deliberate plan (Nagel 1961, 325). Imagine for example a gardener, preparing the ground of his orchard before planting a tree, who uncovers a chest of jewels. Saying that the gardener found the jewels "by chance" signifies "the gardener found a highly desirable object while pursuing a completely different objective." In this context, the reference to chance is understood against a background of purposiveness. This version of the concept of chance is the oldest of all. It is the one Aristotle comments upon in his *Physics* (II, 5). In this well-known text, Aristotle explains that something occurs through "fortune" (τύχη : *tyche*) or "chance" (τὸ αυτόματον : *to automaton*) when a given effect accidentally occurs during "events for the sake of something," in other words, when an end is achieved without having been the cause of the accomplishment of the effect (Aristotle 2008). In English, the best ways of expressing this idea are "luck" and "bad luck." As we will see, this concept of chance, which is so useful in the fields of psychology and history, can be freed from its intentional aspect, and plays an extremely important role in evolutionary theory.

3.1.2 Random Events

The word "chance" takes on a more technical meaning when it is applied to events for which we do not know the causes, that is, the determinant conditions – the antecedents that lawfully determine the course of events. This sense is more recent, and can be explained more precisely. We say that an event is random when we know that events occur as a function of certain classes of defined conditions, but when we do not know which particular conditions are involved in a given situation. We mean that there are several possibilities, but that our knowledge of the event does not enable us to predict which will take place. For example, when the roulette ball falls into a given slot, we have every reason to believe that this event depends on certain physical conditions, the nature and effects of which could be predicted in a determinist manner by a calculator-observer of infinite capacity. The notion of chance in the sense of "random" is more stringent than that of "luck." It requires that we have a hypothesis about what is random, and that we are able to show that we are indeed in a random situation. The only satisfactory solution that has been found to this problem of circularity is the calculation of probabilities, which breaks the vicious circle through an operational solution. A random event is therefore by definition an event that obeys a law of probability, freeing us of the notion of cause. An event that obeys a law of probability can be the result of a perfectly determinist causal process. This is the case in roulette, or in dice games. But a law of probability does not necessarily imply that such causality exists. Quantum mechanics is a classic example of this situation.

3.1.3 Contingency with Respect to a Theoretical System

I will begin by an example of the third sense of "chance." Let us imagine a consistent, idealized theoretical system of "classical mechanics," that is, more or less, Newton's laws of motion plus a certain number of mathematical tools such as differential calculus. In this theoretical context, we can in principle infer the position of a planet in the solar system at any point in the past or the future, on condition that we know the mass, position and speed of each of the components of the solar system at a given moment in time. To make such a prediction, it is clear that it is not enough to know Newton's laws (or any more sophisticated version of classical mechanics). Apart from statements that have the value of a law, statements with regard to the initial conditions are also required. These initial conditions (that is, the parameters that describe the real state of the solar system at a given moment) are said to be contingent with respect to the theoretical system of classical mechanics. Here the notion of contingency takes on a precise operational meaning. It is not a question of saying that an event or a class of events are in and of themselves contingent. The same element can be contingent in one theoretical context, and not in another. For example, in the context of Galilean physics, the value of g, the coefficient of acceleration, is contingent with respect to its theoretical system (that is, the law of falling bodies), because it can be determined only theoretically. Galileo's theoretical system does not make it possible to fix this value. But in the context of Newton's physics, the value of g is no longer contingent. It can be deduced if sufficient information is available with regard to the mass and form of the Earth. On the other hand, although the mass of the Earth is a contingent element in the Newtonian system, its form is not, as it is a consequence of the system that the Earth is spherical with flattened poles.

This example suggests a third possible sense of the term "chance" in the natural sciences. Some classes of events can be called fortuitous to the extent that they are not predictable within the framework of a given theory, either because the theory does not allow the prediction of these events, whatever the available empirical information, or because we do not know with sufficient precision the initial conditions that would make it possible to successfully apply the theory, or again because the calculations involved are too complex.

The definitions of the various concepts of chance I have just outlined are not original (see Ayer 1969; Nagel 1961). Other concepts of chance exist, in the work of these and other authors, which I have not considered here (see Gayon 1994), in particular that of chance as the meeting of independent causal series (Cournot 1984, § 40 [1843]). The three notions of chance presented here seem to me to be sufficient to analyze the way that contemporary biologists make use of chance in their doctrines.

3.2 Modalities of Chance in the Biology of Evolution

In modern evolutionary theory, the notion of chance has a fundamental importance at several levels: mutation, random genetic drift, genetic revolutions, ecosystems and macro-evolutionary phenomena.

3.2.1 Mutation

From a Darwinian perspective, the advantages provided by a mutation are independent of the physical causes that were responsible for the mutation event. These causes are most often electromagnetic rays, chemical factors, viral transposition or transduction events that alter the ordinary functioning of genetic recombination. These factors explain the mutation event, but not the fact that it is favourable. When a Darwinian says that mutations are random, this is only from the point of view of the advantage or disadvantage that they provide in a given environment. The mutation is therefore fortuitous "in the sense that the probability that it occurs is not affected by its virtual usefulness" (Futuyama 1979, 249). This fundamental doctrine has existed ever since Darwin, give or take some changes in vocabulary (Darwin did not speak of mutations but of "variations").

The concept of chance that is involved here is related to that which I characterized earlier by the word "luck." The occurrence of an advantageous (or disadvantageous) mutation can be compared to what happens to the gardener who accidentally finds jewellery while digging his garden. The gardener did not find the jewellery because he was looking for it, but this discovery could have very important consequences for him. In the case of genetic mutations, the conceptual situation is comparable, with the difference that here we do not need to refer to any intentional factor. Individual organisms do not have a physiology that enables them to make precise favourable mutations; at best they can sometimes control the rate of mutation. But once a mutation has appeared, it can have functional consequences that are sufficiently important to affect the individual's chance of survival and reproduction.

We should finally note that if a mutation is seen recurrently in a population, it takes on a form that is determinist, and no longer fortuitous from the point of view of the evolution of this population. However, this is another problem altogether.

3.2.2 Random Genetic Drift

This process is defined at the level of a locus in a Mendelian population. Real populations are always finite, even if they are sometimes extremely large. As a result, there is always a sampling effect in the distribution of gene frequencies at each generation. Genetic drift is a phenomenon that requires a relatively sophisticated mathematical analysis. Nevertheless, a simple example can outline the principles involved. Consider a gene that exists in two allelic forms, in a population that consists of only two individuals, a male and a female. Both individuals are heterozygous (Aa) for this locus. If the population size remains stable, that is, if these two individuals produce two offspring, the probability that they have two homozygous offspring of the same type (either AA or aa) is 1/8[1]. There is thus quite a high probability that one of the alleles could be lost in a single generation. Only a few generations would be required for one of the two alleles to become fixed in the population. In a large population, the chances of losing an allele in a single generation are much smaller, but in the long term they are not infinitely small. The theory of random genetic drift predicts the effects of random sampling in a population of a given size over a large number of generations. Genetic drift is, we say, a factor of stochastic evolution. The basic idea is that the frequency distribution of genes in a given generation is never fixed in a purely determinist fashion (by factors such as recurrent mutations, selection or migration), but must always be described in terms of a law of probability.

The meaning of "chance" in random genetic drift is the same as that involved in roulette or in a dice game. It is a form of chance which we can represent only through a law of probability. It is now recognized that this factor of evolution is extremely important at the molecular level. In fact, it is now accepted that it is the factor of evolution that is responsible for most cases of allelic fixation at the molecular level (chromosomal nucleotide sequences and protein amino acid sequences). We should also note that Motoo Kimura's "neutral mutation – random drift hypothesis of molecular evolution"[2] states that evolution at the molecular level is largely neutral. Note that this idea of neutrality only applies to genes *at a given moment*. A nucleotide substitution in a gene can be neutral in one evolutionary context, and not be so in another.

3.2.3 Genetic Revolution

This idea, introduced in 1954 by Ernst Mayr, is linked to the notion of random drift, but is different, and it also has a different meaning from the point of view of chance. A genetic revolution is a consequence of genetic drift at the level of the genome. When a population is rapidly and severely reduced in size, as often occurs when an island is populated by a species, or in a peripheral, isolated population, some loci are fixed in a homozygous state, because of genetic drift. As a result, the genetic environment of many other genes is altered, and the selective values of some alleles are modified. This situation is relatively well known in experimental population genetics, since the study carried out by Georges Teissier in 1947 on variation in the frequency of the *ebony* gene in a stationary population of *Drosophila*. This study was carried out at a time when the expression "genetic revolution" did not yet exist, but it can be seen retrospectively to have provided an experimental illustration of this idea (Teissier 1947). Many other cases have been described since. But genetic revolutions, while they may be extremely probable in natural populations, are very difficult to detect in these populations.

As Maxime Lamotte (1988) has shown, the type of chance that is involved in genetic revolutions is very different from that associated with random drift. Genetic revolutions are undoubtedly provoked by random drift events, but their fortuitous character does not have the same meaning. Lack of predictability is not due to the stochastic character of the phenomenon, but to the complexity of the interaction of genes, in a given environment, and at the level of a population. These interactions are probabilistic, and they result in a selective process, which is itself deterministic. But, the process is too complex or too poorly known in practice, and the biologist is not able to predict the evolution of the system with the theories (genetical, physiological, populational) at his or her disposal. The notion of chance that is involved here is therefore that which I described earlier under the title "contingency with respect to a theoretical system." For a population geneticist, the physiological effects of the genetic interactions in a given organism, and the selective values of the genotypes that occur in a given environment, are contingent pieces of information in terms of the relatively idealized and simplified theoretical system that they use. As a theory, population genetics provides barely more than a description of the kinetics of the diffusion of genes. This does not mean that the interaction of genes with each other and with the external environment cannot be studied experimentally. But these phenomena are complex, and highly variable from case to case. The information that the biologist gathers about them is barely integrated into an overall theory which would make it possible to deduce these interactions, rather than merely noting their existence.

3.2.4 The Ecosystem Level

At this level, the epistemological situation is similar to that of genetic revolutions. Evolutionary theory is not only interested in the internal evolution of populations, but also in interactions between populations from different species (trophic relations, competition, parasitism, cooperation, etc.), and in interactions between these groups and aspects of the physical environment (climatic and geographical changes, etc.). At this level, too, there are fortuitous effects that have often been highlighted by specialists in ecology. Maxime Lamotte has spoken of "third-order founder effects" to describe the fortuitous effects that occur at the level of biocoenoses and ecosystems, random genetic drift being a first-order effect and genetic revolutions "second-order effects" (Lamotte 1988). The third order of fortuitous effects are, of course, superimposed on the second- and first-order effects, but they cannot be reduced to them. At the synecological level, the complexity of the processes involved in changes to the equilibrium of the flora and fauna is generally beyond the predictive power of available ecological models. This is a similar epistemological situation to that of genetic revolutions. At this level, "chance" means that the available theoretical models are under-determined relative to the available experimental data.

3.2.5 The Macroevolutionary Level (Paleobiology)

Stephen Jay Gould underlined the importance of the "contingency" of evolution on a geological scale (Gould 1989). However, the massive success of this formulation does not mean that it is completely rigorous. I think that those paleobiologists and evolutionary biologists who have adopted the thesis of the large-scale contingency of the history of life have mixed up two ideas, which can be separated in the light of the concepts of chance that I have outlined above.

When paleobiologists speak of the contingency of the history of life, they mean that this history, if replayed, would probably take a different course, and yet we would still use the same general explanatory hypotheses, in particular Darwinian schemas, to account for these different histories. This flows from the importance and the complexity of the initial conditions involved in evolutionary explanations. If this epistemological situation is viewed optimistically, it can be said, as a certain number of philosophers of biology have said (including myself – Gayon 1993, 2005) that biology is fundamentally an historical science. That is, it is a science in which the ultimate explanatory schemes are those of historical causality rather than nomological schemes. In other words, in the final analysis everything boils down to chains of irredeemably singular events. But it can also be said that this situation shows the extent to which paleontological data is under-determined by the available theoretical explanations, which would imply that the theories are not sufficiently powerful to account for the large classes of phenomena they have to deal with. These phenomena are therefore "contingent" in terms of our available theories. They are as they are. Once again we encounter the third meaning of "chance" that we have already referred to several times. The facts outstrip the explanatory capacity of the available theories.

Nevertheless, paleobiologists sometimes have in mind a somewhat different idea when they speak of contingency at the macroevolutionary level. One of the major arguments used by those paleobiologists who have contested the neo-Darwinian orthodoxy has been that of mass extinctions (Raup, Sepkoski, Jablonski). In mass extinctions, they say, the unfortunate species that die out do not do so because they are less fit than their competitors in an adaptive race in response to given ecological conditions, which is the standard Darwinian explanation of extinctions. For example, if the diatoms survived better than other planktonic forms during the mass extinctions of the Cretaceous, it is because they had the luck to possess a favourable trait (the ability to encyst), which had evolved for reasons which were, no doubt, not linked to the physical conditions that led to the mass extinctions. Adapted to become dormant during polar winters, the diatoms found themselves, by chance, with an advantage during a sudden cooling of the Earth; they succeeded only because of an accidental occurrence, and not because they won the race to adaptation through natural selection.

The meaning of "contingency" invoked here is simply the first meaning of "chance" given above, that is, "luck." In other words, the concept of chance operative here is strictly the same as that invoked when it is said, in an orthodox Darwinian context, that mutations occur by chance. The scale of the phenomenon is different, that is all. The word "luck" in fact appears explicitly in the title of David Raup's book on mass extinctions: *Extinction: Bad Genes or Bad Luck?* (Raup 1991).

3.2.6 Other Cases

In this paper I have dealt only with those chance evolutionary phenomena that are solidly based in the modern evolutionary literature. However, there are two new fields of research that are becoming increasingly important, and which both relate to relatively high levels of integration of evolutionary phenomena.

One relates to the dynamics of species diversification. In a recent book, Stephen Hubbell (2001) has presented a stochastic model of the genesis of biodiversity. According to this model, which the author characterizes as a "null hypothesis," the probabilities of the appearance and extinction of a species are likely to be the same for species in competition with each other. The advantage of this model is that it enables us to compare the levels of biodiversity in different habitats. This approach to the origins of biodiversity is an alternative to that based on competitive exclusion, which has dominated ecology since Lotka, and was subsequently adopted by many other writers. The notion of chance involved here is the same as in the case of random genetic drift.

The other field of research that I would like briefly to highlight is that of the chaotic evolution of a number of biological systems. In many fields, ranging from theoretical population genetics and theoretical ecology to the modelling of paleobiological phenomena, a significant place is now given to linear models. These models give a high degree of unpredictability to phenomena, either because modelling them involves equations that have different solutions depending on minuscule changes to initial conditions, or because they do not and cannot have a solution (a situation familiar to mathematicians since the work of Poincaré). If it turned out that such models are appropriate in many evolutionary situations, this would involve a different notion of chance from those outlined above.

3.3 Conclusion

In conclusion, it is legitimate to pose two questions. On the one hand, are the various modalities of chance that have been identified in evolutionist discourse truly distinct, or can they be reduced to a single concept? On the other hand, is this a subjective chance, that is, a chance that exists due to the limits of our knowledge, or an objective chance? These two questions are linked. At first sight, all the forms of chance discussed here appear to be related to subjective chance: the luck of the gardener or the advantageous mutation is only a chance in terms of our limited knowledge of the whole chain of causes and effects. Similarly, the stochastic chance of random drift is only a chance because we have only an overall, statistical vision of this phenomenon. The random drift studied by geneticists is not a radical chance like that of quantum mechanics, it is a "chance" that is the result of our inability to follow in detail the whole series of physically determinist events that are involved in the production of the gametes. Finally, the idea of contingency with respect to a theoretical system is an explicit recognition of the limited capacity of the available theoretical models.

If all these "chances" are subjective, it is tempting to say that they refer definitively to various situations of unpredictability, each of which illustrates the limits of our knowledge.

Nevertheless, it is possible to see things from another philosophical point of view. I want to highlight the conception of chance put forward by Antoine-Augustin Cournot in the nineteenth century. For Cournot, chance should be interpreted as an objective notion, telling us something about the world, and not merely about our power of understanding. This is how we should interpret Cournot's famous statement that chance is a combination of independent causal series. For Cournot, this definition was linked to his rejection of Laplace's universal determinism. Cournot rejected the idea of an order to the world, which was synoptically given once and for all through the fiction of a perfectly informed calculator. He preferred the idea that there exist objective orders of the world, partially distinct, that causality should be seen in terms of these partially isolated and partially connected worlds. Bertrand Saint-Sernin, in his commentary on Cournot, has elegantly described this solution as follows: "the universe has a real history, [and] was created through a series of events none of which was the necessary consequence of the events that preceded it and none of which contained a series of necessary consequences" (Saint-Sernin 1998, 88 – *my translation*).

I hope I will be forgiven this final dialectical twist. There is always a moment when, in science and in philosophy, argument has to give way to decision. The extreme importance of the notion of chance in evolution, with its subtle varieties, clearly shows the historical character of evolution and provides one of the best illustrations of Cournot's vision of the world as composed of partially isolated and partially connected systems at all levels of description.

Notes

1. The probability for the first child to be AA is ¼. The probability for the second child to be also AA is ¼. Since these two events are independent, the probability of having two AA children is $\frac{1}{4} \times \frac{1}{4} = 1/16$. Similarly, the probability of having two aa children is 1/16. Therefore the probability of having either two AA or two aa children is $1/16 \times 1/16 = 1/8$.
2. This hypothesis was first presented in Kimura (1968), but had been previously presented at the Congress of Genetics held in Fukuoka, in November 1967 (cf. Kimura 1983, 28). The expression "neutral mutation – random drift hypothesis of molecular evolution" was first used in Kimura and Ohta (1973, 19) to describe retrospectively the ideas developed in Kimura (1968).

Acknowledgements I thank Bernard Chevassus-au-Louis and Denis Couvet for their fruitful comments on the manuscript. I am indebted to Matthew Cobb for his helpful linguistic corrections.

References

Aristotle. 2008. *Physics*. Trans. R. Waterfield, introduction and notes D. Bostock. Oxford: Oxford University Press.

Ayer, A.J. 1969. *Metaphysics and common sense*. London: Macmillan.

Cournot, A.-A. 1843/1984. *Exposition de la théorie des chances et des probabilités*. Paris: Vrin. 1st ed., published by Librairie de L. Hachette, Paris.

Futuyama, D.J. 1979. *Evolutionary biology*. Sunderland: Sinauer Associates.

Gayon, J. 1993. La biologie entre loi et histoire. *Philosophie* 38: 30–57.

Gayon, J. 1994. Le hasard dans la théorie évolutionniste moderne. *Bulletin de la société d'histoire et d'épistémologie des sciences de la vie* 1: 4–16.

Gayon, J. 2005. De la biologie comme science historique. *Les Temps Modernes* 60e année, (mars-juin, 630–631): 55–67.

Gould, S.J. 1989. *Wonderful life. The Burgess shale and the nature of history*. London: Hutchinson Radius.

Hubbell, S.P. 2001. *The unified neutral theory of biodiversity and biogeography*. Princeton: Princeton University Press.

Kimura, M. 1968. Evolutionary rate at the molecular level. *Nature* 217: 624–626.

Kimura, M. 1983. *The neutral theory of molecular evolution*. Cambridge: Cambridge University Press.

Kimura, M., and T. Ohta. 1973. Mutation and evolution at the molecular level. *Genetics* 73(suppl): 19–35.

Lamotte, M. 1988. Phénomènes fortuits et evolution ? In *L'évolution dans sa réalité et ses diverses modalités*, ed. M. Marois, 243–268. Paris: Masson.

Mayr, E. 1954. Change of genetic environment and evolution. In *Evolution as a process*, eds. J.S. Huxley, A.C. Hardy, and E.B. Ford, 157–180. London: Allen and Unwin.

Nagel, E. 1961. *The structure of science. Problems in the logic of scientific explanation*. London: Routledge and Kegan Paul.

Raup, D.M. 1991. *Extinction: Bad genes or bad luck?* New York: W.W. Norton and Company.

Saint-Sernin, B. 1998. *Cournot*. Paris: Vrin.

Teissier, G. 1947. Variation de la fréquence du gène *ebony* dans une population stationnaire de Drosophiles. *Comptes rendus hebdomadaires des séances de l'Académie des Sciences* 224: 1788–1789.

Chapter 4
Some Conceptions of Time in Ecology

Jean-Marc Drouin

Abstract Whether one is dealing with variations in the size of populations, changes in landscapes, or modifications in the composition of species, all these phenomena are characterized by their temporal structures. Although, like geology, ecology is a historical science, it is also a science of processes like physiology. It is in the combination of these two aspects, and by using both of these paradigms, that the present paper looks for the conceptions of time specific to ecology. Thus, overall representations of ecological phenomena have brought several conceptions of time into play, which can be distinguished in terms of the timescale, its rhythm and its structure. Schematically, descriptions of ecological processes have been founded successively on the idea of a cycle, then on the idea of organic growth, before coming around to unpredictability and chaos. At a more detailed level, this succession of paradigms goes hand in hand with the continued use of concepts that were characteristic of a previous paradigm. The success of some classical concepts can thus be measured by their ability to be inscribed into a new theoretical framework.

There was a time when time moved backwards. The motion of the universe was the reverse of what it is today. Men emerged out of the earth, began their life with old age and became younger from adulthood to childhood before disappearing into nothingness. This is the myth proposed by Plato in *The Statesman*, which features many other elements besides this reversal of temporality (Plato 1925, 268d–274e). Plato imagines that during this epoch the general course of the world was under the control of the Deity, while lower deities ruled over each region and took care of each species, including humans, who at that time did not need any political

J.-M. Drouin (✉)
Department Men-Natures-Societies, Centre Alexandre Koyré,
Muséum National d'Histoire Naturelle, 75005, Paris, France
e-mail: jmdrouin@wanadoo.fr

D. Bergandi (ed.), *The Structural Links between Ecology, Evolution and Ethics: The Virtuous Epistemic Circle*, Boston Studies in the Philosophy of Science 296,
DOI 10.1007/978-94-007-5067-8_4,

constitution. But this time has passed, and the statesman must not be seen as a divine shepherd. It is this conclusion that serves to place this myth in the argumentation of the dialogue. Nevertheless, it is significant that Plato explains that in this world where "all the fruits of the earth sprang up of their own accord for men," the creatures did not "eat one another" (*Ibid.* 271d–271e). Using fiction, Plato emphasizes the link between the temporal processes involving living beings and the complexity of the interactions between them.

Thus Plato offers us a myth that epitomizes the important role of time in ecology. Whether one is dealing with variations in the size of populations, changes in landscapes, or modifications in the composition of species, all these phenomena are characterized by their temporal structures. The present contribution is intended to relate some of these characteristic structures and the conceptions of time that have been used when studying them across the history of ecology (see Egerton 2000; Acot 1988; Deléage 1991; Drouin 1991). The French phrase "figures du temps" serves well as an umbrella term covering all these conceptions, and it has been used as the title of a comprehensive book, edited by Lambros Couloubaritsis and Jean-Jacques Wunenburger and published in 1997. It was also employed in the title of Ivar Ekeland's 1984 essay, "Computing and the unpredictable. Conceptions of time from Kepler to Thom" (*Le Calcul, l'imprévu. Les figures du temps de Kepler à Thom*). The French word "figure" emphasizes the structure – linear or circular, predictable or unpredictable – rather than the nature – objective or subjective – of our conceptions of time.

4.1 Scales of Time

The most striking feature of the different conceptions of time is their different scales, as has been rightly stressed in a collective work from 2000 edited by Monique Barrué-Pastor and Georges Bertrand, *Les temps de l'environnement*. Indeed, Jean-Marie Legay states the issue in his introductory chapter: "When we think of the environment, we spontaneously situate the present in a window of a few decades in the past and a few years in the future. It will require a great effort of reflection and sophistication to escape from this window, and a great deal of persuasion to lead the general public out of it" (Legay 2000, 26 – *my translation*). In another contribution to the same book, Claude and Georges Bertrand remark that one commonly speaks of resources as renewable or non-renewable, but without specifying the period for the renewal (Bertrand and Bertrand 2000, 71). This same issue arises in many debates concerning the evaluation of the impact of human activity on other living species, with the contrasting optimistic and pessimistic views being underpinned by different perceptions of temporal scales. A notion such as the reduction of biodiversity has a meaning only if a particular period of time has been specified. In the controversy concerning endangered species one can easily argue that the extinction of species is inevitable and inherent to the evolutionary process, but this argument can be countered by recalling that what is frightening about today's extinctions is the acceleration in the pace of the process and the increasing number of species affected. Thus, one cannot discuss the problem of extinction without determining the rate at which the process is occurring, which means in turn determining a timescale for considering the problem.

Such problems of timescale are only too well-known to historians. In one of his writings on historiography, his masterwork on the Mediterranean, Fernand Braudel suggests "to divide historical time into geographical time, social time and individual time" (Braudel 1996, 21 [1949]). Individual time is the time associated with the history of events, a short history full of dangers and illusions. Social time is the time of social groups and covers their evolution over the middle term, while geographical time is a timescale that reveals "a history whose passage is almost imperceptible, that of man in his relationship to the environment" (*Ibid.* 20). Scholars do not, however, all agree with the image of an essentially invariant relationship between societies and their environment that is often presented. Those who study the environments of lakes, for example, very often deal with physical entities that formed at the end of the last Ice Age, and so are more recent than the presence of *Homo sapiens* in certain regions (Bertola et al. 1999). What remains pertinent in Braudel's point of view, however, is the way he situates the events of human history on the surface of a slower and deeper history, which in turn is inscribed onto the background of a long-term history. This principle is at the heart of the process of periodization.

4.2 The Chronological Issue

A similar approach can also be found in geology. The history of science stresses how the dismissal of a short chronology of the Earth based on a literal interpretation of the Bible was tied to the development of geology. Regardless of whether contemporary authors resisted this revolution in the conception of time or enthusiastically embraced it, they did perceive its importance. Thus, for instance, in 1844 Ralph Waldo Emerson wrote in his diary that, "The use of geology has been to wont the mind to a new chronology," and he added that geology provides us with a clock even if it is "a coarse kitchen clock" compared to the clocks provided by astronomy (Emerson 2003, 102–103). The witticism of the American philosopher vividly reminds us that early geologists were able to establish only an order of succession among events, the dates of which remained unknown, thereby enabling them to compile only a relative chronology. Nowadays, the methods of isotope-dating allow geologists to establish absolute chronologies. This progress does not, however, render obsolete the division of time into era, periods, epochs, and stages. At a given level, the succession of categories, for instance the periods, depends on a structure, which is named a partition by the logicians. In a paper published in 2004 in the journal *Géodiversitas*, René Zaragüeta-Bagils, Hervé Lelièvre and Pascal Tassy suggest replacing the partition by a hierarchical classification in which each new period would be included in the preceding one, which in turn would itself be included in the one that precedes it. For instance the Cretaceous would be included in the Jurassic, the Jurassic would be included in the Triassic, and so on. According to Bagils, Lelièvre and Tassy, this suggestion – despite its paradoxical appearance – has the advantage of presenting the temporal data in a logical structure comparable to that of phylogenetic trees (for a discussion of the conception of time underpinning such a proposition, see Zaragüeta-Bagils and Bourbon 2005). While the decision concerning whether to adopt or reject this innovation belongs to the palaeontologists and the systematicists, one can nevertheless raise the issue of whether such a formal representation of time could be used in ecology. Its adoption in this context seems unlikely, however, since ecological phenomena are not framed in a unique chronology that would at the same time be unique and specific to ecology. Although a historical science like geology, ecology is also a science of processes like physiology. It is in the combination of these two aspects, and by using both of these different paradigms, that the present paper looks for the conceptions of time specific to ecology (see Drouin 1991, 1994).

4.3 Crop Rotation

Using "paradigm" in the sense given to this term by Thomas Kuhn (1962), one can define the paradigms of scientific ecology as the successive and competing frameworks in which observations have been interpreted and theoretical constructions have been built up. First, we can note that for a long time these paradigms were characterized by their conceptions of time. These can be sorted into two groups: one based on the idea of a cycle and the other on the idea of growth.

Even before the word "ecology" was coined by Ernst Haeckel in 1866, nineteenth-century botanical geography had deployed a cyclical conception of time for the description of phenomena. This conception can be retrospectively considered as ecological, although no such term appears in the founding texts of botanical geography (Drouin 2008, 174–178). Linnaeus devoted some of his academic essays to the idea of the balance of nature (see Linné 1972; for a thorough history of the concept of the balance of nature see Egerton 1973). In these texts, he portrays a providentially organized world, which has no history other than that of the regular increase of habitable land and the preservation of an initial order through the struggle of all against all. Less than half a century later, Alexander von Humboldt, in a lecture to the Institut National on plant geography, incited botanists to study the distribution of plants on the surface of the globe and to recount their migrations (Humboldt 1807, reprinted in Acot 1998, 19–50). For this, Humboldt believed that it was necessary not only to study fossils but also to investigate human history. Humboldt proposed three timescales for this plant geography: first, the short-term timescale of physiological processes on which the physical factors of the environment could act; second, the timescale of human activity; and third, the long-term timescale of geological periods with their catastrophes and their imperceptible changes.

When we look for a conception of a timescale specific to ecology – in the sense of a dynamic inherent to ecological phenomena – we can find it in the works of Dureau de La Malle, a French man of letters. Author of erudite studies of Roman antiquity, Dureau de La Malle devoted several dissertations to the history of domesticated animals and cultivated plants. A landowner in Perche (in the department of the Orne in Normandy), he drew on botanical observations made on his lands to prepare a paper that he read at the Academy of Science on 1 September 1824, which was published in the *Annales des sciences naturelles* in 1825 under a long title that can be translated: "Dissertation on alternation, or on the problem of whether the alternate succession in the reproduction of plant species living in societies is a general law of nature?" (De La Malle 1825, 353–381, partly reprinted in Acot 1998, 117–131). Dureau de La Malle mentions the beneficial effect of the rotation of crops before going on to describe the topography and the nature of the soil on his estate. He then reports the phenomenon of spontaneous succession of species that he has observed in the woods as well as in the meadows. Thus, for instance, as soon as one clears an area in an oak forest featuring a few other trees, the soil is immediately covered with scrub, foxgloves, whortleberries, heathers, birches, and aspens. After a second mowing of the area, the birches and aspens return to the area. Then, after the third clearing of the area, some 90 years later, the oaks and the beeches reconquer the field and remain its absolute masters from then on. Dureau de La Malle reports other cases such as the alternation of whortleberries and heathers in a clearing and the alternation of grasses (Graminae) and white clover (*Trifolium repens* L.) on a lawn in Paris. In conclusion, Dureau de La Malle states that the theory of alternation, which is the "foundation of any good agriculture," can be extended to every plant and can be considered as "a fundamental law imposed on vegetation by the author of all that exists" (1825, 381; 1998, 131 – *my translation*).

A similar idea is expressed in the works of Henri Lecoq, who makes the same link between botany and agriculture (see Drouin and Fox 1999; Pénicaud 2002). This pharmacist, who taught natural history in Clermont-Ferrand, was well known in his time for his studies on botanical geography. He was also the author of several practical books, including one on plant hybridization, which was referred to by Mendel, and a treatise on fodder plants published in 1844 under the French title of *Traité des plantes fourragères*. He devoted several pages of this book (545–549) to the question of alternation that had exercised Dureau de La Malle. Though Lecoq's object was the natural rotation observed in permanent meadows, he nevertheless invoked an example taken from the botany of a forest. Comparing "the successive development of all these vegetables" to a "succession of crops conceived by a skilled agronomist," he concluded that "alternation is so present in Nature that it soon led farmers to the rotation of crops" (Lecoq 1844, 545–549 – *my translation*). And to demonstrate that the Romans already knew this "great law of nature," Lecoq quotes a verse from Virgil: "thus also, with changes of crop, the land finds rest" (Virgil 1999, *Georgics*, I, 82, 104–105).[1] Independently of Dureau de La Malle and Lecoq, the American philosopher and naturalist Henry David Thoreau, in 1860, in a paper entitled, "The succession of forest trees," evokes, albeit implicitly, crop rotation when he writes: "if the wood was old, the sprouts will be feeble or entirely fail; to say nothing about the soil being, in a measure, exhausted for this kind of crop" (see Worster 1985, 70 and following pages; Egerton and Walls 1997).

While in retrospect one can see this first conception of the ecological succession of species as being derived from the rotation of crops, for the authors who formulated this, it was the other way around: they considered the natural phenomenon to be a model for the agricultural practice. Be that as it may, the cycles that are under consideration in this case do not imply an authentic return to the original state of affairs. When the leaves grow again, the tree is one year older, and when the oak comes back, the forest is a century older. The circularity of time in this conception is only approximate, and a helical form might well provide a more appropriate description. Nevertheless, we can consider it to be a circular conception of time if we think initially only of the repeating temporal sequence of change, and then, by extension, apply this pattern to the vision of time itself. Moreover, alternating succession is the principal conception of a dynamic inherent to vegetation, and, unlike many other temporal processes, this dynamic does not imply any irreversibility. The cycle can reproduce itself indefinitely without any degradation of the system. Evidence for this is provided by the way in which Dureau de La Malle thinks of the extension of the pattern that he described to the vegetation of tropical countries where "the extreme variety of species gathered and mixed by nature in the same field is a form of permanent alternation" (Dureau de La Malle 1825, 353, reprinted in Acot 1998, 117 – *my translation*). If the "extreme variety" is similar to a "permanent alternation," alternate succession can conversely be reduced to a form of diversity stretched out in time. Here, the incipient domain of plant ecology tacitly makes use of the conception of time formulated by Leibniz a century earlier, when he defined Space as "the order of coexistences" and Time as "the order of successive existences" (letter from Leibniz to Conti, 6 December 1715, quoted in Robinet 1957, 42). The shortcomings of this parallel between time and space are obvious: while one can freely travel around space in any direction, and every movement can be cancelled out by a reverse movement, time is characterized by its irreversibility. At the end of the nineteenth century, the founders of the field of thermodynamics sought to take this irreversibility into account. This integration of the principle of irreversibility was first realized within the physical sciences, but its effects were also felt in the biological sciences. In contrast to the natural tendency to lose energy as formalized in thermodynamics, the phenomena of growth and evolution appeared to be so many sources of organization and novelty, the opposite of the increase in entropy that characterized other physical processes. Henri Bergson built up a philosophy based on this principle, considering life to be "an effort to re-mount the incline that matter descends" (Bergson 1911, 245 [1907]). Ecology did not escape this movement, as can be seen from the theoretical development of the concept of succession (Lepart and Escarre 1983).

4.4 Succession and Equilibrium

The pioneer in this field was Henry Chandler Cowles. In a study on the dunes of Lake Michigan published in 1899, Cowles described how over the course of a few decades the beach is changed into a moving dune and how in turn this wandering dune is changed into a static dune and how, finally, the beach can pass from this static dune into a forest stage (Cowles 1899). Cowles termed this stage the normal "climax" of the region. Although it constituted a novelty in this context, the term "climax" was not a neologism. Originally a Greek word, climax had been integrated into English in order to name the peak of a progressive process.

During this same period, another American naturalist, Frederick Edward Clements, made this concept of climax the key to his theory of vegetation. This theory likens the plant community to an organism, and the phenomenon of ecological succession to a process of growth. Clements expresses this idea dramatically: "As an organism the formation arises, grows, matures and dies." Extending the analogy, he writes: "Succession is the process of the reproduction of a formation," and he adds, "this reproductive process can no more fail to terminate in the adult form of vegetation than it can in the case of the individual plants" (Clements 1916, 124–125; see Egerton 1973, 344). Several ecologists supported this vision of ecological change, which was widely discussed. Nevertheless, this discussion did give rise to competing theories, reinterpretations, and opposing conceptions.

Among the competing theories was one proposed by William Skinner Cooper in 1926. Cooper's theory explicitly rests on a different conception of change and therefore a different conception of time. Cooper states that "a sound conception of the fundamentals of dynamic ecology must be based upon the premise of the universality of change." In this view the vegetation is "presented as a flowing braided stream" that disappears and reappears, with more or less separate and definite elements, branches, interweavings, and anastomoses (Cooper 1926, 397–398). This valuable analogy allows a new interpretation of the concept of climax.

For Cooper, the climax is one of the large slowly-moving currents of the braided stream, formed by the merging of many streamlets. It occurs when all change-inducing factors are acting with extreme slowness. It comes into being insensibly. In a given spot, however, it usually terminates abruptly through the agency of some sudden acting factor, the result of which event will be the forking of the large stream, illustrated by the initiation of one or more "secondary successions" (*Ibid.* 409).

The British botanist Arthur George Tansley proposed another reinterpretation of Clements's theory in a paper published in 1935 and entitled, "The use and abuse of vegetational concepts and terms." This paper is often cited, as it is the place where the term "ecosystem" was originally coined. Tansley stresses that the ecosystem includes a complex not only of organisms but also of inorganic factors, and he describes it as a "particular category among the physical systems that make up the universe." The climax can be considered as a "relatively stable dynamic equilibrium" and as an instance "of the universal processes tending towards the creation of such equilibrated systems" (Tansley 1935, 306). Tansley also states that: "There is in fact a kind of natural selection of incipient systems, and those which can attain the most stable equilibrium survive the longest," and he adds two explicit philosophical references: "A corresponding idea was fully worked out by Hume and even stated by Lucretius" (Tansley 1935, 300. See Lucretius 1982, book V, 418–508, 411–417; Hume 1991, part VIII, 143–147).

Nearly 10 years before the publication of Tansley's paper, an American botanist called Henry Allan Gleason launched another criticism of Clements's theory. Rejecting the analogy of the plant community as a super-organism, Gleason proposed the fundamental principle that "every species is a law unto itself." It grows where it has been disseminated by chance and where it can grow. It "grows in company with any other species of similar environmental requirements." From this perspective, the succession of plants is nothing other than the change of vegetation that one can observe when there is a change in one of the primary factors responsible for the introduction or elimination of species, migration, or environmental selection. As for the climax, it represents "a stage at which effective changes have ceased, although their resumption at any future time may again initiate a new series of succession" (Gleason 1926, 25–26).

Despite the great differences between them, the views of Clements and Gleason on temporal change can nevertheless be analyzed using the same mathematical model, as was demonstrated by Michael B. Usher in a paper published in 1981 in the journal *Vegetatio*. In this paper, Usher characterizes Clements's position as being "deterministic" and Gleason's as being "stochastic." He stresses how Markovian models adopt Clements's view, "relying on the fact that if succession is an orderly process then probabilities from one state to another can be estimated," and he stresses too how the same models fit with Gleason's view. Thus, in so far as each individual plant is "an entity in its own right," it is possible "to estimate a series of probabilities which define all of the possible outcomes for the fate of one individual" (Usher 1981, 12. Concerning Markovian models in ecology, see also Usher 1979). Here, ecology comes close to the ideas expressed by Ilya Prigogine and Isabelle Stengers in a book whose title can be translated *Between Time and Eternity*: what one calls "Markov chains" do not result from a purely random process nor from a deterministic algorithm. In the first case, the knowledge of the start of the sequence leaves the continuation completely indeterminate. In the second case, knowledge of initial conditions could allow one to predict the continuation. In the case of the Markov chain, every term that can follow a given term or a given sequence of terms is characterized by a probability (Prigogine and Stengers 1992, 89).

Asking the question "can we imagine a natural mechanism producing such a chain?" the authors answer in the affirmative, and propose as an example a mechanism of chemical reactions. Ecology could have provided another example to support their view.

4.5 Irreversibility and Unpredictability

The application of mathematical models to ecological processes is nothing new, however, as they have been used for a long time in the study of population dynamics. In the decade between 1925 and 1935, Alfred James Lotka in the United States and Vito Volterra in Italy built up a mathematical description of the variation of size in animal populations, although quite independently of one another (on the history of population dynamics, see Kingsland 1985; Israel and Gasca 2002). It was in this context that they formulated the equations that still bear their names. These equations describe two related cyclical fluctuations, one for the number of prey, and the other for the number of predators. Thus, to model the growth of a population in an environment with a limited capacity to sustain this population, Lotka as well as Volterra developed a now well-known formula, which, when the size of the population is plotted on the y-axis, and time along the x-axis, generates a characteristic S-shaped curve. Thus, one can now see two classical images of temporal change, the repeating cycle and the one-time growth curve, two contrasting and complementary images, which both promise to be able to predict the future state of a system (on unpredictability in ecology and what it implies concerning the place of man in nature, see Blandin and Bergandi 2000; Bergandi 2007).

This idealized model requires putting the temporal characteristics of the individuals in question to one side, at least temporarily. Thus, in 1935, when Volterra and Umberto d'Ancona ask the reader to suppose that the individuals of each species are identical, it is because they intend neither to deal with any differences in age, size, sex, and so on, nor to take into account the periodicity arising out of births and deaths (Volterra and d'Ancona 1935, 14). Recognizing this form of abstraction makes it easy to understand the interest of the work of Patrick Leslie in this area (Leslie 1945; for an explanation, see Lacroix 1987; Begon and Mortimer 1986, 54–60; on the history of their discovery, see Caswell 1989, 24–26). The population matrix named after him, the "Leslie matrix", factors the numbers of individuals in different age-groups, the fecundities associated with different ages, and the age-specific survival rates into the computation of the changing size of a population. Proposed in 1945, this method took advantage of the development of computing power after the war and is still widely used. Thus, the Leslie matrix provides a model that allows a finer-grained analysis than the simple S-shaped curve model, but which similarly offers the hope of providing predictability. But this promise of predictability associated with the ideals of the classical physical sciences has come to see itself threatened by the extension of chaos theory to ecology (concerning the history of the theory of chaos, see Ekeland 1984; Gleick 1987; Boutot 1993; Dahan 2000; Aubin and Dahan 2002).

In 1974, Robert M. May, an Australian-born scholar, trained in physics and working in the department of biology at Princeton University, published a paper in *Science* entitled "Biological populations with nonoverlapping generations: stable points, stable cycles, and chaos." In the first lines of the paper, May distinguishes between two kinds of biological populations: those where growth in the population is a "continuous process" (e.g. the human being) and those where this growth "takes place at discrete intervals of time" (e.g. some species of insects). In the former type of populations, generations "overlap," while in the latter, they do not. In the case of nonoverlapping generations, the number N_{t+1} of the population at a time $t + 1$ is obtained by using the formula $N_{t+1} = N_t\,[1 + r\,(1 - N_t/K)]$, with r being "the usual growth rate" and K the "carrying capacity" of the environment. Robert May stresses that if this simple model predicts a stable equilibrium for $0 < r < 2$, it gives rise to "essentially arbitrary dynamical behaviour" "once r becomes large enough" ($r > 2.570$). Referring to Edward Lorenz's seminal works, May writes: "Such behavior has previously been noted in a meteorological context, and doubtless has other application elsewhere. For population biology in general, and for temperate zone insects in particular, the implication is that even if the natural world were 100% predictable, the dynamics of population could nonetheless in some circumstances be indistinguishable from chaos, if the intrinsic growth rate r were large enough" (May 1974, 645).

The fact that the size of a population becomes stable for some values of the growth rate while becoming chaotic for other values that lie very near the first set of values makes this a good example for introducing students to chaos theory. Indeed, at the end of another paper, May himself emphasizes that the "most important applications" of this model "may be pedagogical" (May 1976, 467). He is convinced that even "in the everyday world of politics and economics, we would be better off if more people realised that simple nonlinear systems do not necessarily possess simple dynamical properties" (*Ibid.*).

Thus, one can apply what Prigogine and Stengers have written on chaotic systems in general to the ecological model studied by May: they free unpredictability from the contingency of ignorance and give it an intrinsic meaning (Prigogine and Stengers 1992, 81). Furthermore, it is in such chaotic dynamics that Prigogine and Stengers see the possibility of building this bridge which Boltzmann did not succeed in building between dynamics and the world of irreversible processes (Prigogine and Stengers 1992, 107).

This view expressed by the authors of *Between Time and Eternity* concerning the statistical interpretation of entropy – which they link to the name of Ludwig Boltzmann – converges with the analysis made, half a century earlier, by Lotka in his 1925 book, *Elements of Physical Biology*:

> The failure of the differential equations of dynamics to discriminate between t and $-t$ raises the question as to the physical significance and origin of our subjective conviction of a fundamental difference between the forward and the backward direction in time – a conviction that is intimately bound up with the concept of evolution, for, whatever may ultimately be found to be the law of evolution, it is plain that no trend of any kind can be defined or even described without reference to a favored direction in time (Lotka 1925, 37–38).

As Lotka himself summarized his position a decade later in his French publication *Théorie analytique des associations biologiques*, it is difficult to speak of progressive changes, not only because there is no objective definition of progress, but also because we do not have any objective criterion for establishing the direction of time. Thus, we are obliged to content ourselves with our subjective judgment of the dissymmetry in the course of a time (Lotka 1934, 30–31). Lotka's reflections on time did not receive any great echo in their own epoch. Lotka's lack of recognition was resented by the biologist and mathematician Vladimir Kostitzin, who wrote in a letter to Volterra from 31 December 1935 that Lotka's thoughts on the absence of any contradiction between the existence of living matter and the law of entropy have not been sufficiently appreciated by biologists and philosophers of nature (Israel and Gasca 2002, 233).

4.6 Persistence and Anticipation

Thus, overall, the representations of ecological phenomena have brought several conceptions of time into play, which can be distinguished in terms of the timescale, its rhythm, and its structure. As these theoretical constructions have their own history, they must be inscribed in a temporality that brings out the innovation and radical novelty of many scientific theories as well as the persistence of certain themes. Schematically, the descriptions of ecological processes have been founded successively on the idea of cycle, then on the idea of organic growth, before coming around to unpredictability and chaos. At a more detailed level, this succession of paradigms goes hand in hand with the continued use of concepts that were characteristic of a previous paradigm. Thus, the success of some classical concepts can be measured by their ability to be inscribed into a new theoretical framework.

Perhaps most surprising of all, however, is the presence of themes in ancient texts that look like anticipations of analogous modern themes. We can, for example, cite the verses of the *Georgics* where Virgil compares the repeated selection of seeds by the peasant to the action of an oarsman fighting against the stream. This can be read as a representation – originating in agronomy – of the man's active resistance against the increase in disorder in the universe (see Virgil 1999, *Georgics*, I, 197–203, 112–113).[2] It is perhaps ironic, then, that reflecting on the notion of time in ecology can lead to the propagation of anachronisms, albeit in moderation.

Notes

1. "sic quoque mutatis requiescunt fetibus arva" (Virgil, *Georgics*, I, 82, 104–105).
2. "Vidi lecta diu et multo spectata labore
   degenerare tamen, ni vis humana quotannis
   maxima quaeque manu legeret. Sic omnia fatis
   in peius ruere ac retro sublapsa referri,
   non aliter quam qui adverso vix flumine lembum
   remigiis subigit, si bracchia forte remisit
   atque illum in praeceps prono rapit alveus amni."

   "I have seen seeds, though picked long and tested with much pains, yet degenerate, if human toil, year after year, culled not the largest by hand. Thus by law of fate all things speed towards the worse and slipping away fall back; even as if one, whose oars can scarce force his skiff against the stream, should by chance slacken his arms, and lo! headlong down the current the channel sweeps it away" (Virgil, *Georgics*, I, 197–203, 112–113).

Acknowledgements I would like to thank Jonathan Simon for correcting my English.

References

Acot, P. 1988. *Histoire de l'écologie*. Paris: Presses Universitaires de France.

Acot, P. (ed.). 1998. *The European origins of scientific ecology*, 2 vols. Amsterdam: Overseas Publishers Association/Gordon and Breach Publishers/Editions des Archives Contemporaines.

Aubin, D., and A. Dahan Dalmedico. 2002. Writing the history of dynamical systems and chaos: *Longue durée* and revolution, disciplines and cultures. *Historia Mathematica* 29: 273–339.

Barrué-Pastor, M., and G. Bertrand (eds.). 2000. *Les temps de l'environnement*. Toulouse: Presses Universitaires du Mirail.

Begon, M., and M. Mortimer. 1986. *Population ecology. A unified study of animals and plants*. Oxford: Blackwell Scientific Publications.

Bergandi, D. 2007. Niveaux d'organisation: Evolution, écologie et transaction. In *Le tout et les parties dans les systèmes naturels*, ed. T. Martin, 47–55. Paris: Vuibert.

Bergson, H. 1907/1911. *Creative evolution*. Trans. A. Mitchell. New York: Henry Holt and Company. 1st French ed., published by Les Presses universitaires de France, Paris.

Bertola, C., C. Goumand, and J.-F. Rubin (eds.). 1999. *Découvrir le Léman, 100 ans après François-Alphonse Forel*. Geneva: Slatkine, Musée du Léman.

Bertrand, C., and G. Bertand. 2000. Le géosystème: un espace-temps anthropisé. Esquisse d'une temporalité environnementale. In *Les temps de l'environnement*, eds. M. Barrué-Pastor and G. Bertand, 65–76. Toulouse: Presses Universitaires du Mirail.

Blandin, P., and D. Bergandi. 2000. L'homme et la nature: les sciences changent de role. In *Nature vive*, ed. C. Larrère, 88–99. Paris: Muséum National d'Histoire Naturelle, Nathan.

Boutot, A. 1993. *L'invention des formes. Chaos, catastrophes, fractales, attracteurs étranges et structures dissipatives*. Paris: Odile Jacob.

Braudel, F. 1949/1996. *The Mediterranean and the Mediterranean world in the age of Philip II*. Trans. S. Reynolds, 2 vols. Berkeley: University of California Press. 1st French ed., published by A. Colin, Paris.

Caswell, H. 1989. *Matrix population models: Construction, analysis and interpretation*. Sunderland: Sinauer Associates.

Clements, F.E. 1916. *Plant succession: An analysis of the development of vegetation*. Washington, DC: Carnegie Institution of Washington.

Cooper, W.S. 1926. The fundamentals of vegetational change. *Ecology* 7: 391–413.

Couloubaritsis, L., and J.-J. Wunenburger (eds.). 1997. *Les figures du temps*. Strasbourg: Presses Universitaires de Strasbourg.

Cowles, H.C. 1899. The ecological relations of the vegetation on the sand dunes of Lake Michigan. *Botanical Gazette* 27: 95–117, 167–202, 281–308.

Dahan Dalmedico, A. 2000. L'image 'fin de siècle' des sciences. La théorie du chaos a-t-elle engendré une révolution scientifique ? *La Recherche* 327: 58–61.

Deléage, J.-P. 1991. *Une histoire de l'écologie*. Paris: Editions du Seuil.

Drouin, J.-M. 1991/1993. *Réinventer la nature. L'écologie et son histoire*. Paris: Desclée de Brouwer. Repr., *L'écologie et son histoire. Réinventer la nature*. Paris: Flammarion.

Drouin, J.-M. 1994. Histoire et écologie végétale: les origines du concept de succession. *Ecologie* 25: 147–155.

Drouin, J.-M. 2008. *L'herbier des philosophes*. Paris: Editions du Seuil.

Drouin, J.-M., and R. Fox. 1999. Corolles et crinolines: le mélange des genres dans l'œuvre d'Henri Lecoq. *Revue de Synthèse* 4: 581–599.

Dureau de La Malle, A. 1825. Mémoire sur l'alternance ou sur ce problème: la succession alternative dans la reproduction des espèces végétales vivant en société, est-elle une loi générale de la nature? *Annales des sciences naturelles* 5: 353–381.

Egerton, F.N. 1973. Changing concepts of the balance of nature. *The Quarterly Review of Biology* 48: 322–350.

Egerton, F.N. 2000. Ecology. In *Reader's guide to the history of science*, ed. A. Hessenbruch, 190–192. London/Chicago: Fitzroy Dearborn Publishers.

Egerton, F.N., and L.D. Walls. 1997. Rethinking Thoreau and the history of American Ecology. *The Concord Saunterer* 5(new series):5–20
Ekeland, I. 1984. *Le calcul, l'imprévu. Les figures du temps de Kepler à Thom*. Paris: Editions du Seuil.
Emerson, R.W. 2003. *Selected writings*. With an introduction by C. Johnson. New York: Signet Classic, Penguin.
Gleason, H.A. 1926. The individualistic concept of the plant association. *Bulletin of the Torrey Botanical Club* 53(1): 7–26.
Gleick, J. 1987. *Chaos*. New York: Viking.
Hume, D. 1991. In *Dialogues concerning natural religion*, ed. S. Tweyman. London: Routledge.
Israel, G., and A.M. Gasca. 2002. *The biology of numbers. The correspondence of Vito Volterra on mathematical biology*. Basel: Birkhaüser Verlag.
Kingsland, S.E. 1985. *Modelling nature. Episodes in the history of population ecology*. Chicago: University of Chicago Press.
Kuhn, T. 1962. *The structure of scientific revolutions*. Chicago: Chicago University Press.
Lacroix, H.J. 1987. Tables de survie et matrices de Leslie. In *Une approche mathématique de la biologie*, ed. R.V. Jean, 39–70. Chicoutimi: Gaëtan Morin.
Lecoq, H. 1844. *Traité des plantes fourragères ou Flore des prairies naturelles et artificielles de la France*. Paris: H. Cousin Libraire-Éditeur.
Legay, J.-M. 2000. Les temps de l'environnement. In *Les temps de l'environnement*, eds. M. Barrué-Pastor and G. Bertrand, 19–32. Toulouse: Presses Universitaires du Mirail.
Lepart, J., and J. Escarre. 1983. La succession végétale, mécanismes et modèles: analyse bibliographique. *Bulletin d'Ecologie* 14(3): 133–178.
Leslie, P.H. 1945. On the use of matrices in certain population mathematics. *Biometrika* 33: 183–212.
Linné, C. 1972. *L'équilibre de la nature*. With an introduction by C. Limoges and trans. by B. Jasmin. Paris: J. Vrin.
Lotka, A.J. 1925/1956. *Elements of physical biology*. Baltimore: Williams & Wilkins. Repr., *Elements of mathematical biology*. New York: Dover.
Lotka, A.J. 1934. *Théorie analytique des associations biologiques. Première partie. Principes*. Paris: Hermann.
Lucretius. 1982. *De rerum natura*. Trans. W.H.D. Rouse, rev. M.F. Smith. Cambridge, MA: Harvard University Press.
May, R.M. 1974. Biological populations with non-overlapping generations: Stable points, stable cycles, and chaos. *Science* 186: 645–647.
May, R.M. 1976. Simple mathematical models with very complicated dynamics. *Nature* 261: 459–467.
Pénicaud, P. 2002. *Henri Lecoq. Les fortunes d'un naturaliste à Clermont-Ferrand*. Clermont-Ferrand: Mémoires de l'Académie des sciences, belles lettres et arts de Clermont-Ferrand.
Plato. 1925. *The Statesman. Philebus. Ion*. Trans. H.N. Fowler and W.R.M. Lamb. Cambridge, MA: Harvard University Press.
Prigogine, I., and I. Stengers. 1992. *Entre le temps et l'éternité*. Paris: Flammarion.
Robinet, A. 1957. In *Correspondance Leibniz-Clarke*, ed. A. Robinet. Paris: Presses Universitaires de France.
Tansley, A.G. 1935. The use and abuse of vegetational concepts and terms. *Ecology* 16(3): 284–307.
Thoreau, H.D. 1860/1906. The succession of forest trees. In *The writings of Henry David Thoreau*, vol. V, 184–204. Boston: Houghton, Mifflin. An address read to the Middlesex Agricultural Society in Concord, Sept. 1860.
Usher, M.B. 1979. Markovian approaches to ecological succession. *Journal of Animal Ecology* 48: 413–426.
Usher, M.B. 1981. Modelling ecological succession, with particular reference to Markovian models. *Vegetatio* 46: 11–18.

Virgil. 1916/1999. *Eclogues, Georgics, Æneid I-VI.* Trans. H.R. Fairclough, rev. G.P. Goold. Cambridge, MA: Harvard University Press.

Volterra, V., and U. D'Ancona. 1935. *Les associations biologiques au point de vue mathématique*. Paris: Hermann.

von Humboldt, A. 1807. *Essai sur la géographie des plantes*. Paris: Levrault.

Worster, D. 1985. *Nature's economy: A history of ecological ideas.* Cambridge/New York: Cambridge University Press.

Zaragüeta-Bagils, R., and E. Bourdon. 2005. Pithecanthropus erectus. *Sciences et Avenir* 142(hors-série): 58–63.

Zaragüeta-Bagils, R., H. Lelièvre, and P. Tassy. 2004. Temporal paralogy, cladograms and the quality of the fossil record. *Geodiversitas* 26(3): 381–389.

Chapter 5
Facts, Values, and Analogies: A Darwinian Approach to Environmental Choice

Bryan G. Norton

Abstract Most writing on environmental ethics concerns the dichotomy between humans and non-humans, and much of the work in the field has been motivated by the effort to escape "anthropocentrism" with respect to environmental *values*. Resulting debates about whether to extend "moral considerability" to various elements of non-human nature have been, to say the least, inconclusive. In this paper, a new approach to re-conceptualizing our responsibilities toward nature is proposed, an approach that begins with a re-examination of spatio-temporal scaling in the conceptualization of environmental problems and human responses to them.

5.1 Introduction

Most writing on environmental ethics concerns the dichotomy between humans and non-humans, and much of the work in the field has been motivated by the effort to escape "anthropocentrism" with respect to environmental *values*. Resulting debates about whether to extend "moral considerability" to various elements of non-human nature have been, to say the least, inconclusive. In this paper, a new approach to re-conceptualizing our responsibilities toward nature is proposed, an approach that begins with a re-examination of spatio-temporal scaling in the conceptualization of environmental problems and human responses to them. Before ending this introductory section with a more practical definition of this kind of management – sometimes called "adaptive" – I will briefly explain how the new approach rests on four intellectual pillars:

B.G. Norton (✉)
Philosophy, Science, and Technology Program, School of Public Policy,
Georgia Institute of Technology, Atlanta, GA 30332, USA
e-mail: bnorton@gatech.edu

D. Bergandi (ed.), *The Structural Links between Ecology, Evolution and Ethics: The Virtuous Epistemic Circle*, Boston Studies in the Philosophy of Science 296,
DOI 10.1007/978-94-007-5067-8_5,

I. *A Commitment to a Unified Method: Naturalism.* Attempts to separate factual from value content in the process of deliberation are rejected; there is only one method for evaluating human assertions, including assertions with all kinds of mixes of descriptive and prescriptive content, and that is the method of experience – active experimentation, when possible, and careful observation otherwise. The scientific method is embraced as the best approach to evaluating hypotheses about cause and effect, but also about what is valuable to individuals and cultures.

II. *An Empirical Hypothesis.* The values of people who care about the environment are expressed in (a) the ways that they "bound" the natural system associated with a given problem, and (b) the choices that they make in focusing on the physical dynamics that they use to "model" those problems.

III. *A New Approach to Scaling and Environmental Problems.* Building on this Empirical Hypothesis, scalar choices in modeling environmental problems can be explicitly treated as expressions of the values of the residents of a place. What is considered important by people who inhabit a place provides one kind of evidence that can guide model-builders in choosing a temporal horizon over which impacts will be measured, and processes of change monitored, and these values can also guide the "modeling" of dynamics and problem-drivers. This approach to evaluating change can be understood as a search for a locally-anchored conception of sustainability and sustainable development.

IV. *A Darwinian Approach to Knowledge, Value, and Managerial Choice.* "Adaptive Management" builds upon a Darwinian analogy. It applies the Darwinian mechanism of survival, analogically, to human community groups, which are interpreted as socially-evolved groups with abilities to adapt, culturally, to an environment. Since cultural adaptation includes the inheritance of acquired characteristics, the velocity of adaptation is greatly increased over genetic evolution, and this necessitates our re-calibration of our "horizons of concern." A rational reproducer, knowing that his/her offspring will succeed in the long run only if a livable environment is perpetuated, must "manage" the environment for a longer scale. A livable environment includes enough options and opportunities for offspring and cohorts to continue a way of life, which supports a *schematic definition* of "sustainability."

This paper, exploring this Darwinian analogy, considers the implications of "managing adaptively" in response to problems faced by real communities, members of which respond from the viewpoint of their sense of place and space. It will be shown that careful use of this analogy can provide a fertile way of thinking about both environmental "facts" and "environmental values."

This approach to management is often referred to as "adaptive management" in North America,[1] but it is practiced elsewhere in varied forms and with different names. This approach is Darwinian in a specific sense to be explained here, and it sets out to use science and social learning as tools to achieve cooperation in the pursuit of management goals. In the United States, the ideas were first articulated by the scientific and philosophical forester, Aldo Leopold (1949), who emphasized the importance of multi-scalar adaptation in his essay, *Thinking like a Mountain*, and who advocated for scientific management throughout his career. The centrality of notions of scale to Leopold's understanding of management is emphasized in my book, *Sustainability: A Philosophy of Adaptive Ecosystem Management* (2005), where I make the case that Leopold was the first "adaptive manager," though the term was not in use in his time.

I define adaptive management as management efforts that share three characteristics:

1. Experimentalism: adaptive managers respond to uncertainty by undertaking reversible actions and studying outcomes to reduce uncertainty at the next decision point.
2. Multi-Scalar Modeling: adaptive managers model environmental problems within multi-scaled ("hierarchical") space-time systems.
3. Place-Orientation: adaptive managers address environmental problems from a "place," which means that problems are embedded in a local context of natural systems but also of political forces.

By profession, most adaptive managers are ecologists and most discussions to date have emphasized learning our way out of scientific uncertainty; these ecologists have paid less attention to developing appropriate processes for *evaluating environmental change* and for setting intelligent *goals* for environmental management. In this paper I will propose a theory of values and valuation that explains and supports an adaptive approach to management and extends this approach to include social learning about values and goals. Finally, I will explain in which sense it can be considered Darwinian and adaptive. Each of the following four parts constitutes a somewhat more detailed explanation of Pillars I–IV, above.

5.2 Naturalism: The Method of Experience

As noted above, much discussion in environmental ethics has centered on the debate between anthropocentrists and non-anthropocentrists, between those who limit moral considerability to humans and those who extend human considerability into the non-human world. Unfortunately, environmental ethicists have not paid as much attention to another controversial dichotomy, that between "facts" and "values" – between descriptive and prescriptive language. Analytic philosophers have been very cautious about mixing facts and values in argumentation, a trend initiated by David Hume, and intensified by George Edward Moore, who argued that "good" and similar normative terms cannot be defined in naturalistic terms because they express "non-natural qualities." Moore cautioned against committing the "naturalistic fallacy," which he took to involve defining normative terms such as "good" in terms of observable characteristics. Recently, two prominent environmental ethicists have, adopting arguments reminiscent of Hume and Moore, argued for forsaking science and descriptive studies and concentrating on "intrinsic values" in the effort to protect natural systems, processes, and elements.

In particular, J. Baird Callicott (2002) and Mark Sagoff (2004) have argued that environmentalists should play down instrumental arguments for saving species and biodiversity, basing their main arguments on the "intrinsic value" of Nature. Sagoff says: "indeed environmental policy is most characterized by the opposition between instrumental values and aesthetic and moral judgments and convictions" (2004, 20). He goes on to argue that, "Environmental controversies ... turn on the discovery and acceptance of moral and aesthetic judgments as facts" (*Ibid.* 39). Unfortunately, he describes no means to separate fact from fiction in assertions that this or that has intrinsic value and explicitly claims that scientific arguments have no bearing on defending environmental values or goals.

Callicott agrees with Sagoff in sharply separating science from ethics and instrumental uses from non-instrumental appreciation: "We subjects value objects in one or both of at least two ways – instrumentally or intrinsically – between which there is no middle term" (2002, 16). Callicott goes on to emphasize the subjective source of these intrinsic values: "All value, in short, is of subjective provenance. And I hold that intrinsic value should be defined negatively, in contradistinction to instrumental value, as the value of something that is left over when all its instrumental value has been subtracted. In other words, 'intrinsic value' and 'non-instrumental value' are two names for the same thing" (*Ibid.* 21). Emphasizing this point, he says: "Indeed, it is logically possible to value intrinsically anything under the sun – an old worn out shoe, for example" (*Ibid.* 10). Callicott and Sagoff, then, relying on a sharp dichotomy between descriptive and prescriptive discourse, and on sharply separating instrumental reasons for protecting nature from non-instrumental reasons, have thus called for a strategy of emphasizing intrinsic values over instrumental uses of nature in arguing for the protection of nature. These non-natural qualities are, apparently, apprehended through intuition or created by emotional affects, and they seem ill-suited to provide inter-subjectively valid or convincing reasons for environmental action. Worse, they tend to isolate discussions of environmental values from environmental science (Norton 2008, 2009).

What concerns me about this trend is that it isolates environmental valuation and goal-setting from scientific discourse, creating ideological positions *about the nature of value*, polarized positions that cannot be resolved by experience or deliberation. To place this argument in historical context, Sagoff and Callicott seem to be reviving arguments against "naturalism" in the study of environmental values and goals, relying on mysterious "non-natural" qualities – qualities not visible or measurable by the senses.

Philosophically, Hume's prohibition, sometimes called "Hume's Law," which cautions against deriving "ought statements" from "is statements," may be taken narrowly, to recommend against trying to "deduce" prescriptions from descriptions; if so taken, it does not condemn the methods developed here, because the proposed methods do not use deduction, but rather employ an open-ended deliberative process that encourages transformations in both beliefs and in value commitments. Similarly, I take Moore to rule out the definition of values in terms of facts; but even if that is a "fallacy," one can adopt alternative views of naturalism that define much looser, but still effective, rational connections between scientific information and the articulation and defense of values.

My position follows Bernard A.O. Williams (1985), who argued persuasively that, in ordinary discourse, fact discourse and value discourse are inseparable; when philosophers separate them, they do so on the basis of a specialized theory, such as logical positivism. In the ordinary discourse in which citizens discuss and evaluate their environment, these discourses are inseparable; to insist on partitioning the two discourses is to artificialize policy discourse. There is an alternative, of course. Following pragmatists such as Charles Sanders Peirce and John Dewey, I advocate a pragmatic epistemology for environmental science and policy discourse. This epistemology insists upon a single method – the method of experience – and I apply this method to both factual claims and evaluative ones. Like Dewey, I take assertions that some thing or some process is *valued* to be a hypothesis that that thing or process is *valuable*. Pursuing that value, and acting upon associated values, provides communities with experience that can support or undermine the claim that the thing or process is indeed valuable.

Speaking practically, I fear that non-naturalism, whereby intrinsic values are linked to subjective feeling or to intuition, will not provide good reasons to adopt environmental goals. How can environmental advocates rely upon intrinsic values if they are as ephemeral and idiosyncratic as commitments to old shoes? But this is only the beginning of the problems. Insisting that environmentalists express their environmental values in non-natural qualities, non-naturalists divorce environmental values from environmental sciences, because non-natural environmental values do not translate into measurable physical parameters of environmental systems. I fear that, if environmental ethicists lead environmental advocates away from naturalistic approaches to the environment, the possibilities for integrating environmental science and environmental values will be lost.

So, the first pillar of my proposed approach is a form of methodological naturalism. While it does not contradict Hume (if understood as prohibiting the *deduction* of values from facts), nor need it violate Moore's cautions against *defining* values as facts (in terms of natural qualities), my commitment to naturalism is supported by the method of experience as the sole means to assess and evaluate environmental change. This method, while not expecting *deductions* from facts to values, or definitions of values in terms of facts, relies on the open-ended, public process of challenging beliefs and values with contrary experience. From these challenges, we expect attitudes, values, and beliefs to change – but the changes cannot be justified by deductive arguments flowing one way from facts to values. The changes needed to support a new conservation consciousness are usually reorganizations and re-conceptualizations of facts, not deductions from value-neutral facts. The goal of this approach is to link values held by citizens in a place as closely as possible to natural processes that give that place its distinctiveness – the historical intertwining of natural and cultural processes at a physical location. The specific means by which assertions of value are connected will be through the development and refinement of measurable indicators that reflect values articulated by the stakeholders who represent multiple positions within the community. Choosing multiple indicators, as will be discussed in the next two

Sections, will express a commitment to pluralism; the new approach to scaling in problem formulation is taken up in Sect. 5.4, and is intended to begin the process of refining and integrating the multiple values that communities hold for natural systems.

5.3 An Empirical Hypothesis

I have argued that one can be a naturalist without breaking "Hume's Law," or committing Moore's fallacy; but I would like to challenge two assumptions that Hume made in formulating his law. By stating the law as a prohibition against deriving "ought-copula" sentences from "is-copula" sentences, Hume implied that fact discourse and evaluative discourse could be sharply separated, and that the difference would announce itself syntactically via the evident copula. This, I believe, was Hume's mistake: he implies that value discourse and factual/descriptive discourse are separable, when in fact they are all mixed together in ordinary speech; to separate them artificializes normal discourse in important ways.

This argument, however, raises an inevitable question: How, exactly, *do* values manifest themselves in scientific, descriptive literature, which claims to be "value free" and is apparently "scrubbed" of evaluative language before publication in scientific journals? In order to answer this question, it is useful to follow Funtowicz and Ravetz (1990, 1995) in distinguishing between "curiosity-driven" (discipline-driven) science and "mission-oriented" (problem-driven) science. Authors who place their research in disciplinary journals succeed, to varying degrees, in purging evidence of values from their scientific papers. However, in building and shaping an adaptive management approach, I am more interested in mission-oriented science than in playing "find-the-values" in disciplinary science. Funtowicz and Ravetz argue that, because mission-oriented science often takes place in contexts where stakeholders have different perspectives and interests, scientific models and reports that are taken to bear on management decisions will, in effect, be "peer reviewed" not only by appropriate disciplinary scientists, but also by scientists in different fields, and by interested laypersons. This places a transparency requirement on scientific discourse: if science is to be advanced as a guide to controversial policies, then that science must be explainable – and explained – in ordinary speech that requires no scientific credentials to understand.

When the attention shifts from disciplinary science to mission-oriented science, when scientific questions and hypotheses are formed within a controversial management context, values slip back into the discourse, because participants are proposing policies from their own perspective. So, if we want to find values implicit in scientific work, we should look in the discourse of management science. The values and interests of participants, I will argue, are coded in the choices they make to "model" the problem – to bound the problem spatially, to form a temporal horizon, and to describe a physiology of the system that is considered problematic.

It is now possible to state a general hypothesis regarding the values that are implicit in discussions of mission-oriented, environmental science: Values are often embedded in the choices individuals and groups make when they choose a "mental model" of the problem they are addressing. With this very general hypothesis as guidance, I am working with colleagues from geography and economics to test the more specific hypothesis that the values of individuals and groups are often embodied in the spatio-temporal scales that they attribute to the system they have identified as problematic.

We are undertaking to examine scaling choices in two, very different contexts. One group will examine scaling choices of scientific modelers when they try to capture and illustrate the nature of an environmental problem by providing a model of the system that is affected by the problem. A second group will do open-ended interviews and work with focus groups to determine how stakeholders in active, deliberative, adaptive management processes "scale" the systems to which they attribute an environmental problem. Values of both scientists and other stakeholders, according to this hypothesis, will guide – and be expressed in – the ways that they "bound" the system used to analyze the problem, and in choices regarding the dynamics (physiology) of the problematized system.

A historical example may help to illustrate what is claimed in the hypothesis. Chesapeake Bay, on the East Coast of the US, is among the most productive – and loved – bodies of water in the world. The Bay is the mouth of the mighty Susqhehanna River, and many other tributaries that drain a huge portion of the northeastern United States. By the 1970s there were multiple danger signals that the Bay was becoming polluted, even if it was unclear what was driving the widespread changes in the Bay's functioning, especially the increasing turbidity and consequent dieback of the vast underwater grass flats that formed the base of the Bay's food-web. Until the late 1970s, when the US Environmental Protection Agency (EPA) undertook a detailed scientific study, pollution issues had mostly centered around toxic and point source pollution problems, including polluting industries and inadequate sewage treatment in a densely packed area of residences, agriculture, and industry. It was learned that, while environmental monitors were paying attention to small-scale, local variables, a large-scale variable associated with a larger-scale dynamic – one driven by the total input of nutrients into the Bay from its tributaries – posed a slower-moving, but more profound threat to the Bay's health. Agricultural and residential run-off of nitrogen and phosphorous was causing increased turbidity, reducing submerged aquatic vegetation beds, and causing algal blooms and anoxia in deep waters. The rich farmlands of Pennsylvania, the Piedmont, and the coastal plane all drain into the Chesapeake. To save the Chesapeake, it would be necessary to gain the cooperation of countless upstream users of the waters that eventually enter the Bay, a monumental task, since Pennsylvania and the District of Columbia, situated upstream on tributaries, had no coastline on the Bay and no direct stake in its protection.

Nevertheless, against all odds, the larger Bay community – enabled by the EPA study and countless private research efforts – succeeded in transforming the public consciousness to think of the Bay as an organic, connected watershed. Tom Horton, an environmental journalist and activist, said it best when, at the height of this period of intense social learning, he wrote: "We are throwing out our old maps of the Bay. They are outdated not because of shoaling or erosion or political boundary shifts, but because the public needs a radically new perception of North America's greatest estuary" (Horton 1987, 7–8). He pointed out that, as the problem with Bay water quality expanded beyond point-source pollution to include non-point sources, residents of the area had to change their mental model of the processes of pollution; and they had to address activities throughout the watershed, adopting a model that includes all the lands contributing run-off to the Bay.

What I want to draw out of this analysis is that the "transformation" of the Bay from an estuary into a watershed occurred in a context of mission-oriented science, and it was as much a process of transformation of public consciousness as it was a change in scientific understanding. It was a dramatic change in perspective that was driven by values – an outpouring of love and commitments not to let the Bay become more unhealthy. In order to address the problem of Bay water quality, it was necessary to create a new "model" of what was going wrong. The shift in models led to a public campaign, driven by the deep and varied values residents felt toward the Bay, which was marked, for example, by the outstanding success of the Chesapeake Bay Foundation, a private foundation that advocates, educates, and supports science to guide Bay management. So, we have here an example of a value-driven re-conception of the Bay system, how it works, and how pollution is being delivered into it. We can say that a new "cultural model" was formed, and Chesapeake Bay management, while not perfect, of course, has been a model of cross-state cooperation, as serious steps have been taken throughout the watershed to reduce non-point-source as well as point-source pollution. Notice how a shift of models used to characterize the problem corresponds to a shift to new causal models with radically different boundaries and physiology, but the force driving acceptance of these expanded models was the love felt for the Bay. A scientific finding that the Bay was threatened by processes outside its currently-conceived boundaries, interacting with the strength of the love for the Bay as a "place," created a new model that more accurately represented the problems of the Bay, and also expanded the sense of responsibility of its residents and users. Public understanding embraced the larger system, and attention shifted to addressing non-point-source pollution problems throughout the watershed.

So, as my hypothesis would predict, residents and officials of the Bay area, upon being convinced that the Bay's health was threatened, and that a large part of the problem came from the larger-scale watershed system, shifted to a larger perspective on Bay health, a perspective that is perhaps more aligned with a scientific understanding of the problem faced. This shift in perspective, however, is not just scientific: it expresses a deep and varied set of social values that residents and stakeholders feel toward the Bay. And, when Horton describes the change in hydrological and cartographic terms, the underlying truth is that the shift to a watershed-sized model was the expression of an implicit value, a sense of caring for the health of the Bay as a part of one's way of life. This is what I mean by saying that values can be implicit in the choices of "models," scientific and mental, that people use to understand problems. The love and respect residents had for the Bay, once the nature of the threat was better understood, expressed itself in a ready embrace of the Bay as a watershed. Their local valuings came to express a community consensus in goals and values, transforming a local consciousness into a regional consciousness and sense of responsibility. Through social learning, the residents of the area discovered how to "think like a watershed," and began living in a larger "place" than before.

How might one proceed to study community values, values that are thus implicit? (Norton and Noonan 2007). One way might be to study policy-making and public discourse in situations that are not as developed as the three-decades-old watershed management movement in the Chesapeake region. We can compare the progress in the Chesapeake region with management of Lake Lanier, a large impoundment of the Chattahoochee and Flint Rivers at their confluence north of Atlanta, Georgia. Lake Sidney Lanier is a multi-purpose lake providing hydropower, a stable water source for Atlanta, and a tourist and second-home mecca. Here, the story may follow a plot similar to the one in the Chesapeake region. Chicken farming is expanding throughout the hills of the Piedmont, as land in the Lanier watershed gradually gets swallowed up by the northward growth of Atlanta, and the wooded shores of the Lake are replaced with suburban lawns. Phosphorous and nitrogen levels in the Lake are rising, and some scientists fear that expanding anoxia in deep waters will destroy the productive sport fishery.

The society that surrounds the Lake, however, while it is well represented by at least two strong watchdog organizations, has not yet learned to think like a watershed, and the result is run-away sprawl and small municipalities competing for economic growth to help pay taxes to support more and more inefficient support services that are spread too thin. Residents tend to address pollution of the Bay as if the main problem were frequent and unwanted closures of areas to swimming, brief incidents of point-source pollution, caused by outbreaks of bacteria. This means that their "model" of the pollution of the Lake has not been expanded to the watershed as a whole. Our research team will try to identify and articulate any inkling we can find of a nascent sense of place attached to Lake Lanier as a whole. Perhaps the unfolding case can provide a clearer picture of how values, models, scale, and horizon interact to transform perspective and re-align commitments.

5.4 Scaling and Environmental Problem Formulation

Environmental disputes are so difficult, among other reasons, because it is so difficult to provide a definitive problem formulation. This feature was well explained by Rittel and Webber (1973), who distinguished "benign" and "wicked" problems. Benign problems, they said, have determinate answers, and when the answer solution is found, the problem is uncontroversially "solved." Mathematics and some areas of science exemplify benign problems. Wicked problems, on the other hand, resist unified problem formulation; there is controversy regarding what models to use and what data are important. Rittel and Webber suggest that wicked problems, because they are perceived differently by different interest groups with different values and goals, have no determinate solution because there is no agreement on the problem formulation. They can be "resolved" by finding a temporary balance among competing interests and social goals, but as the situation changes, the problem changes and becomes more open-ended. Rittel and Webber explicitly mention that wicked problems have a way of coming back in new forms; as society addresses one symptom or set of symptoms, new symptoms appear, sometimes as an unintended effect of treatments of the original problem.

Most environmental problems are wicked problems; they affect multiple values, and they impact different elements of the community differently, encouraging the development of multiple models of understanding and remedy. While resistance to unified problem formulation is endemic to wicked problems, and requires iterative negotiations to find even temporary resolutions and agreements on actions, one aspect of wicked problems – the temporal open-endedness that often attends wicked problems and brings them back in more virulent form as larger and larger systems are affected – may be susceptible to clarification through modeling. Ecologists have introduced "hierarchy theory" (HT) as a set of conventions to clarify space-time relations in complex systems. HT can be characterized by two axioms (which happen to coincide with the second and third key characteristics of Adaptive Management listed in Sect. 5.1). HT encompasses a set of physical systems that are (a) under study by scientists, and which are (b) attributed two constraints on observer and system behavior: (i) the system is conceived as composed of nested subsystems, such that any subsystem is smaller (by at least one order of magnitude) than the system of which it is a component, and (ii) all observations of the system are taken from a particular perspective within the physical hierarchy. My major addition to HT is to broaden the interpretation of (ii) to (ii'): all observations *and evaluations* are taken from a particular perspective within the physical hierarchy. The effect of this innovation is to make environmental values, evaluation, and social learning about values endogenous to the broader, adaptive management process.

This conceptual apparatus allows us to see human decision makers as located within layered subsystems and super-systems, with the smallest subsystems being the fastest-changing, while the larger systems change more slowly and provide the environment for adaptation by subsystems (including organisms and places – composed of individuals and cultures). This convention allows us to associate temporal "horizons" with changing features of landscapes, as is illustrated in the famous metaphor used by Aldo Leopold, a forest-land manager and a wildlife manager. Leopold set out to remove predators from the Forest Service ranges he managed in the southwestern US. When the deer starved for lack of browse, he regretted his decision to extirpate wolves, chiding himself for not yet having learned to "think like a mountain" (Leopold 1949). He had not yet, that is, understood the role of the targeted species in the broader system. When he came to understand that role, he accepted responsibilities for the long-term consequences of his decisions, and advocated wolf protection in wilderness areas.

Leopold's account corresponds to the above case of Chesapeake Bay. In both cases, human activities – intended to improve the lot of human consumers of nature's bounty – threatened larger-scale dynamics. Thinking like a mountain – or a Bay – requires accepting responsibility – and to extend caring – for what will happen in the future, in a subsequent generation. Accepting this responsibility is just the adoption of a "model" of the world in which a physical system, composed of layers of productive and constraining systems, is being viewed from within that system. At this point of time armed and with some knowledge of changing systems and how to model these, we begin to accept moral responsibility for actions that were once thought to be morally neutral. In both of these cases, accepting moral responsibility – and a sense of caring – was inseparable from adopting a changing causal model of what has happened: to deer populations on Leopold's metaphoric mountain, and to submerged aquatic vegetation in the Chesapeake.

Using this framework of actions embedded within nested, hierarchical systems, it is possible to articulate my new approach to evaluating changes in human-dominated systems. Human management of the environment takes place within environmental systems as they are embedded in larger and larger – and progressively slower-changing – super-systems. Each generation is concerned for its short-term well-being (personal survival), but also must be concerned to leave a viable range of choices for subsequent generations. Given our expanding knowledge of our impacts on the larger and normally slower-changing systems that form our environment, it seems reasonable also to accept responsibility for activities that can change the range of choices that will be open to posterity. We must, that is, think of adaptation as embodying at least two "scales" of time. Not only must each generation replace itself through reproduction, we also know that subsequent generations, like us, will face choices. We can think of our short-term choices as involving "economic" choices – choices about how to use and allocate current resources. These resources are presented to us as a mixture of opportunities and constraints. What Leopold and the residents of Chesapeake Bay learned is that the choices we make today – to maximize economic well-being, for example – when aggregated over our whole population can manifest themselves as changes in the set of opportunities and constraints faced by subsequent generations. For example, Leopold's extirpation of wolves led to an explosion of the deer population, a reduction in vegetative cover, and an increase in erosion and the siltation of streams. Chesapeake residents, busily plying their trades and tending their lawns, discovered that the ways in which they were pursuing their economic well-being could turn the Chesapeake into an anaerobic slime pond. In both cases the total impacts of individual actions to improve individual well-being turn out to reduce the mix of opportunities and constraints faced by subsequent generations. A concept of sustainability nicely "falls out" from this conception of adaptive management, in that a "schematic definition" of sustainability can be constructed on the axioms of adaptive management, provided only that prior generations accept responsibility for their impacts on the choice sets of subsequent generations.

Given this rather sparse set of assumptions and hypothetical premises, it is possible to provide a simple and elegant definition of *sustainability*, or rather what might better be called a definitional schema for sustainability definitions. Because of the place-based emphasis of adaptive management and the recognition of pervasive uncertainty, there is only so much that one can say about what is sustainable at the very general level of a universal definition. Speaking at this level of general theory, sustainability is best thought of as a cluster of variables; local

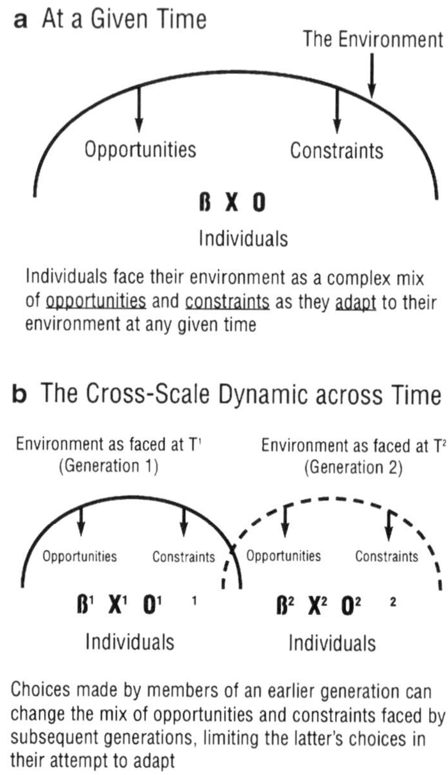

Fig. 5.1 (**a**/**b**). The environment: a complex mix of opportunities and constraints (from Norton 2005)

communities can fill in the blanks, so to speak, to form a set of criteria and goals that reflect their needs and values. Although I recognize the importance of local details in particular determinations of sustainability, the three core characteristics of adaptive management go a long way toward specifying a schematic definition of sustainability. A schematic definition makes evident the *structure* and *internal relationships* that are essential to more specific, locally-applicable definitions of sustainable policies.

First notice that the principles of adaptive management, when embodied in models, place individual actors in a world that is encountered as a mixture of opportunities and constraints; some of the chooser's choices result in survival; the chooser lives to choose again. If the chooser survives and has offspring, the offspring will also choose in the face of similar but changing environmental conditions. Choices of other opportunities lead to death with no offspring. This is the basic structure of an evolution-through-selection model that interprets the environment of a chooser as a mixture of opportunities and constraints; it contextualizes the "game" of adaptation and survival and can be represented as in Fig. 5.1a.

This relationship is simply an expression of the relationship implicit in the second and third principles of adaptive management: the chooser is located at a space-time point within an environmental system, observing and acting from that perspective within the system. The actions of individuals, taken individually and collectively, moreover, can be understood as experiments on two different scales. Survival of the individual depends, in the short run, on very local conditions of stability; but that local stability represents also a negotiation with slower-changing background conditions. The actions, once undertaken, will result in either survival or termination of the individual or the population over varying periods of time. Community-level success, in other words, requires success on two levels: at least some individuals from each generation must be sufficiently adapted to the environment to survive and reproduce, *and* for the population to survive over many generations, the collective actions of the population must be appropriate for (adaptive to) its environment. Since humans are necessarily social animals (because of the long period of helpless infancy of individuals), individual survival depends also on reasonable levels of stability in the "ecological background," the stage on which individuals act. This environment normally changes much more slowly than individual behaviors, permitting adaptation, over generations, to stable aspects of the environment. This simple model, if given a temporal expression, represents the relationship between individuals who live in an earlier generation and those who live later, as represented in Fig. 5.1b.

From this simple framework, a schematic definition of sustainability emerges: Individuals in earlier generations alter their environment, using up some resources, leaving others. If all individuals in the earlier generations overconsume, and if they do not create new opportunities, then they will have changed the environment that subsequent generations encounter, making survival more difficult. A set of behaviors is thus understood as sustainable if and only if its practice in generation *m* will not reduce the mix of opportunities and constraints that will be encountered by individuals in subsequent generations *n, o, p*.

Note that this simple model can be described, in its bare bones, as a natural-selection machine. No value judgments need be implied in the model – it can be viewed simply as representative of the relationship of individual and collective choices as they play out over generations. In each generation, individual actors make their choices *given* extant opportunities; looked at intergenerationally, aggregated choices of individuals may change the mix of opportunities and constraints faced in subsequent generations.

Although the model has a "flat," schematic character, it could also be given a richer, normative-moral interpretation, as is hinted at by use of the terms *opportunities* and *constraints*. If we stipulate that the actors are human individuals, then the simple model provides a representation of intergenerational impacts of decisions regarding resources; our little model can thus be enriched to allow a normative interpretation or analogue. If we accept that having a range of choices is good for free human individuals, we can see the structure, in skeletal form, of the normative theory of sustainability. An action or a policy is not sustainable if it will reduce the mix of opportunities and constraints in the future.

Each generation stands in this asymmetric relationship to subsequent ones: choices made today could, in principle, reduce the range of free choices available to subsequent generations. Thus it makes sense to recognize impacts that play out on multiple, distinct scales. If we can agree that maintaining a constant or expanding set of choices for the future is good, and that imposing crushing constraints on future people is bad, our little model has the potential to represent, and relate to each other, the short- and long-term impacts of choices *and* to allow either a physical, descriptive interpretation or a normative one. This schematic definition, understood within the general model of adaptive management, captures two of our most important basic intuitions about sustainability: that sustainability refers to a relationship between generations existing at different times – a relationship having to do with the physical existence of important resources – and that this relationship has an important normative dimension. Thus we can tentatively put forward adaptive management – complete with a schematic definition of sustainability – as a useful model for environmental science and management.

I have claimed that, *provided one accepts responsibility for one's impacts on the future and the set of choices (adaptations) available to future people*, a plausible definition of sustainability results. In the next Part, I will give a Darwinian explanation of why we "should" accept such responsibility. Before turning to that aspect of the argument, I want first to show how a multi-scalar evaluation of impacts of actions can be correlated with a pluralistic approach to environmental values. If it is recognized that some actions – or aggregations of actions – of individuals threaten a valued aspect of the environment on a multi-generational time scale, there arises a competition between the "good" of current individuals (consumption and increased individual welfare) and the "good" of future people (who we can expect to want to face a broad array of opportunities to adapt to their environment as they see fit), and if we accept that (following Hierarchy Theory) these goods are associated with different social and ecological dynamics, which unfold at different scales, it may be possible to identify public policies that protect both of these dynamics; or, it may be possible to find an acceptable balance among the values if they turn out to be competing. In a pluralistic value system – if it is embedded in a multi-scalar system – some human values can be associated with faster ("economic") processes of production and consumption. The protection of native vegetation and improving Bay water quality, on the other hand, are associated with a large-scaled system and with values that need not compete with economic values, because they unfold at different scales and they are supported by different processes. It becomes conceivable to find a policy that provides an adequate increment of individual welfare, but which does so in a way that does not destroy the long-term value of holding options open for future choosers. This multi-scalar approach to mapping values onto systems and subsystems opens the way for a new approach to environmental values and valuation. Once values discussion and goal-setting are made endogenous to the adaptive management system, we can conceptualize the evaluation of environmental change in a new way, bypassing the question of which beings, objects or processes have a good of their own – the debate that separates economists from philosophers – and evaluate various scenarios or development paths, rather than assessing the value of elements of nature or services derived from nature. A development path is a possible unfolding of the future in a place. Development paths are most useful when they are associated with a proposed strategy of action to move the society toward the outcomes described in the scenario. Development paths can then be assessed according to multiple criteria. It can be fruitful for participants in an adaptive management process to debate which indicators and what goals and targets should be used in measuring managerial success, and to debate the appropriate weightings of the various criteria in making decisions.

This approach can help to illuminate social values through the device of "backcasting" (Robinson et al. 1996). Participants in a goal-setting operation can, first, construct alternative scenarios extending 25 and 50 years into the future. Then, based on these scenarios, goals or "targets" are chosen for future dates, and the question is asked: What should we be doing now to achieve goals set for 25 and 50 years from now?

The evaluation of alternative development paths is now conceived as applying multiple criteria – that are reflective of the love and concern of a people for a place, and also embody the community's commitment to hold open opportunities to maintain a continuous culture. Proposal of, refinement of, and debating about these indicators and how to weight them can *express* the values that a community has, and it can reflect the community's diversity and pluralism in the ways its people value nature. This new approach to studying environmental values shifts the debate from determining which elements of nature have economic, instrumental, or intrinsic value, and focuses attention on choices of what indicators to monitor and what goals to set. In this way, adaptive management "naturalizes" the expression of values, allowing values to be reflected in the choices of what gets monitored and what gets protected, and integrates the discussion of goals and the discussion of science in the search for improved environmental policies.

5.5 Darwin and Environmental Ethics

I have associated, throughout this paper, adaptive management with the ideas of Charles Darwin, a task made easier by C.S. Holling's labeling of the place-based scientific modeling of environmental problems as "adaptive management." One might ask: in what sense is adaptive management reflective of the ideas and models of Darwinian natural selection? Does "adaptive" management come by its label legitimately? Is the application of Darwinian selection to environmental management analogical? Or is adaptive management literally "Darwinian"?

I will argue that adaptive management embodies *both* analogical and literal senses of the idea of adaptation. First, it is useful to recall, as has been pointed out by many biologists and evolutionary theorists, that "selectionism" can be understood as composed of a minimalist model, according to which selection and adaptation are assured by the existence of three simple functions. Provided there is a source of variation in a population (of organisms, societies, or even ideas), and provided there is a means of "coding," by which characteristics are passed from one state to the next, any method of selection will result in selective "survival" and modification of the original variability. This minimal model does not specify the origin of the variability, the exact means of coding, or the mechanisms that sort the population. If we think of selectionism in these stripped-down terms, it is possible to think of the model proposed as a literal application of selectionism to communities and their practices of resource use.

Modeling the survival of human communities, however, cannot be literally Darwinian in a more specific sense, because Darwinian evolution applies at the individual level. Being highly social animals, and highly dependent upon cultural adaptations for survival, human individuals and populations across multiple generations require institutions and practices that are adapted to a place, adapted not just in the short-term sense of flourishing for a few years or decades, but in a long-term sense that the culture must survive through "hundred-year floods," "hundred-year droughts," and many, many cycles of birth and death and birth. So if one thinks of Darwinian selection quite narrowly as individual, genetically coded selection, adaptive management clearly reflects an analogical use of *adaptive*. If, however, one thinks more broadly of group selectionism as occurring among communities, based on learned practices and successful human institutions, then the proposed model suggests that improved management practices – practices that protect the whole range of human values derived from the environment – would literally increase the likelihood of long-term survival of the community by holding open a wide range of options for future adaptations.

Since group adaptation – learning to live for many generations in a place – is cultural and Lamarckian and is based on conscious observation and foresight, it may be possible for cultures to develop adaptations to protect the culture over multiple generations from the loss of opportunities. They can pay close attention to trends and to changes in the landscape as a result of current practices. Conscious learning, based on observation of the effects of existing practices on the ecological system, makes improved fitness possible without a long series of failed experiments. Long-term survival, for adaptive managers and pragmatists, is thus simply a specialized form of social learning – learning to adapt practices and actions to the opportunities and constraints stored in local ecological systems.

Furthermore, we can also appeal to social learning to answer an objection I have often heard raised against the application of Darwinian selectionism to ethical and social issues. Surely, the objection goes, we should aim for more than "survival." Humans, our objector wants to emphasize, care about a lot more than simple physical survival. And surely we do care about more than physical survival. We are emphasizing not mere physical survival, but survival of individuals as *members of an ongoing culture*. The long-term survival of individuals over many generations requires thriving practices and institutions that hold open valuable options for future generations. Survival, thus, cannot be individual physical survival alone; it must include also the survival and thriving of the culture itself. And we know that if the culture is to survive over many generations, it must be intertwined with the development, use, and protection of the land that represents the habitat of the culture. This point can be taken one step further: interactions between individuals, a culture, and the land they inhabit are not only essential for simple survival; they also give meaning to the experiences that are shared by members of the culture. The institutions that sustain the culture must include practices and institutions that embody the stored wisdom of that culture.

Let us complete this brief discussion of whether adaptive management is properly so called by exploring one interesting and provocative analogy that can be borrowed – and modified as necessary – from genetic evolution. This exploration, I hope, will provide some indication of the richness of the structural analogies that can be generated from the Darwinian idea when it is applied provocatively to problems of environmental management.

Consider an analogy originally introduced and developed in the 1930s by Sewall Wright (1932), a giant of early genetics, as an attempt to explain the puzzling role of the new field of genetics in biological evolution. In his venerable model, Wright suggested that in order to set individual adaptation within a broader context of multigenerational evolution of populations, we could understand biological evolution as a game that is played out on an "evolutionary landscape." The long-term adaptive landscape can be conceived as a topological space in which higher ground represents greater fitness, given an environmental situation. In Wright's model, the best strategy for a lineage or a species will be to find a very high peak and climb as high on that peak as possible. In individual-level genetic evolution, which was Wright's topic, however, we assume that individuals have no foresight: they simply apply the rule of finding the steepest incline available to them by pursuing the most attractive opportunity available to them at a given time. In this game the best evolutionary strategy will be to locate the steepest fitness slope that is reachable immediately. If, however, one recognizes that the steepest-sided peak may take one to a low plateau, going up the steepest slope may in fact cause a given lineage to be stalled on a very low fitness plateau. It remains controversial among evolutionary biologists whether it is possible for a lineage, once trapped on a low peak, to go back down and search for a higher peak, but let us assume for the sake of the analogy that such lineages cannot return down the slope to reengage the search for greater fitness opportunities on higher slopes (Kauffman 1995).

Suppose, now, we consider the problem of choosing a sustainable environmental policy as a parallel adaptive game. As in evolution, winning in this game is for members of the group to stay alive to play another generation, and the biggest winners are the ones who keep their markers in the game for the longest time. For individuals, this means surviving long enough to reproduce and make a contribution to the gene pool of a population – and also its culture. But if an individual is a member of a population that goes extinct in a subsequent generation, individual reproductive success will have been futile in the long run. Once evolution is understood as a multi-scalar phenomenon, which affects individuals in the short term and communities in the long term, it follows that organisms or individuals can only "win" on both time scales by also leaving a winnable game board for their offspring. This model explains why, on a Darwinian model, where success for actors is both to survive and to reproduce and also to keep one's offspring in the game as long as possible, a rational actor will care not just about choosing the best opportunity in the present. A rational player in this game will also care about – and take steps to protect – the range of options available to those offspring as they exist in subsequent generations. This concern will include both maintenance of the resources needed to live and maintenance of broader cultural opportunities as well. Holding open opportunities for future generations, in other words, is to exercise foresight with respect to choosing a tall fitness peak for the population/culture of offspring to climb.

To apply Wright's analogy at the level of cultural survival, two important changes would be required. First, as conscious users of symbols, humans have foresight and, unlike other organisms, can, in principle at least, forgo the immediate rewards gained by ascending a steep slope to a low peak if they can see much greater benefits in seeking a higher peak. So our cultural application of Darwinism opens the possibility of balancing a choice between the steepness of slope (short-term edge in competitiveness) and the height of the peak pursued (long-term survival options). Wright's fiction of an unchanging landscape in which organisms try to survive and adapt no longer holds. Adaptation, because of the introduction of consciousness, culture, and complex technologies, will be adaptation to a moving target, a dynamically changing environment in which traits that are adaptive today may become neutral or maladaptive in the future.

If, in fact, humans can affect the adaptive landscape faced by their successors, and if these anthropogenic changes reduce the fitness of future generations of their society or culture, then we could represent this outcome by saying that the action of an earlier generation has reduced the opportunities of future generations to make a living and develop their culture. Certain choices made by one generation might, on this model, reduce the height of the very fitness peak that past generations have already begun to climb. If the anthropogenic impacts are irreversible, analogous to a culture that violently and pervasively changes its habitat, the actions of the earlier generation could be interpreted as trapping its successors on a relatively low peak. The peak that earlier generations chose to climb could in fact be "lowered" if the collective choices of a given generation alter the environment in irreversible and drastic ways.

Wright's analogy, I am suggesting, illustrates the sustainability problem when it is applied at the conscious level at which humans foresee the future and develop technologies to pursue their goals; we can interpret the problem of community adaptability as one of developing a multi-scale strategy against the background of a changing fitness landscape. If human populations can irreversibly damage an environment so as to eliminate choices that would have been open to future generations, then we can characterize the damage to the future as a reduction in fitness. Such a reduction will result from a reduction of the available options to survive and thrive in the place in question. Wright's analogy of the fitness landscape, then, illustrates in Darwinian terms the nature of the intergenerational harm involved in living unsustainably – it involves reducing the options available for future adaptations, by reducing the height of the fitness peak that a culture is climbing. In the worst case, the environment could change so much that future generations are not adapted enough to live; in this case the earlier generation will have directly thwarted its own long-term goals. Or the effect might be mainly cultural and social, affecting the range of opportunities and choices in the future.

Some changes to the landscape/environment may terminate certain of our valued practices of today, resulting in a kind of cultural suicide. People of the future will have nothing in common with us because they face such a changed environment – a different set of opportunities – and will feel no sense of community with us.

Provided one generation in fact cares for its "legacy," it follows that choosers must go beyond an analysis of individual well-being to take into account the range of options they leave for subsequent generations. If we add the goal of long-term survival of our successors (sustainability), then we must have (at least) a two-scale decision process with independent criteria for value at each level. If members of a culture act only on a short-term economic basis, as a group they may find the "steepest peak" of rapid growth while ignoring the goal of multigenerational sustainability. That criterion must tell us how to avoid going steeply up a low peak or, even worse, taking actions today that limit future options (lowering the peak we are climbing). A rational chooser who seeks both individual survival and cultural sustainability would at least set minimal criteria to protect the choices of future generations. I have thus fulfilled the promise that I would provide a Darwinian explanation of why the present does – and should – care about the options open to future people. In addition to information about the short-term economic impacts of our choices, we also need to think about the "height" of the peak that we are choosing, creating, and sustaining. Good choices increase economic well-being without putting us on a lower fitness peak – one that will leave few options available in the future. We must, as Leopold argued, learn to think – and evaluate – like a mountain, on multiple scales of time. I have here suggested an approach that would help us to do so, and to live sustainably.

Note

1. Holling (1978). A note on terminology: perhaps the closest analogue to adaptive management in Europe is "Ecological Modernization," which shares some tenets with adaptive management, but also differs in emphasis. See Hajer (1995).

References

Callicott, J.B. 2002. The pragmatic power and promise of theoretical environmental ethics. *Environmental Values* 11: 3–25.

Funtowicz, S., and J.R. Ravetz. 1990. *Uncertainty and quality in science for policy*. Dordrecht: Kluwer Academic Publishers.

Funtowicz, S., and J.R. Ravetz. 1995. Science for a post-normal age. In *Perspectives on ecological integrity*, eds. L. Westra and J. Lemons, 146–161. Dordrecht: Kluwer Academic Publishers.

Hajer, M.A. 1995. *The politics of environmental discourse: Ecological modernization and the policy process*. Oxford: Oxford University Press.

Holling, C.S. 1978. *Adaptive environmental assessment and management*. London: Wiley.

Horton, T. 1987. *Bay country*. Baltimore: The Johns Hopkins University Press.

Kauffman, S. 1995. *At home in the universe*. Oxford/New York: Oxford University Press.
Leopold, A. 1949. *A Sand County almanac and sketches here and there*. Oxford: Oxford University Press.
Norton, B.G. 2005. *Sustainability: A philosophy of adaptive ecosystem management*. Chicago: University of Chicago Press.
Norton, B.G. 2008. Beyond positivist ecology: Toward an integrated ecological ethics. *Science and Engineering Ethics* 14: 581–592.
Norton, B.G. 2009. Convergence and divergence: The convergence hypothesis twenty years later. In *Nature in common*, ed. B. Minteer, 235–260. Philadelphia: Temple University Press.
Norton, B.G., and D.S. Noonan. 2007. Ecology and valuation: Big changes needed. *Ecological Economics* 64: 664–675.
Rittel, H.W.J., and M.M. Webber. 1973. Dilemmas in a general theory of planning. *Policy Sciences* 4: 155–169.
Robinson, J.B. (ed.). 1996. *Life in 2030. Exploring a sustainable future for Canada*. Vancouver: University of British Columbia Press.
Sagoff, M. 2004. *Price, principle, and the environment*. Cambridge/New York: Cambridge University Press.
Williams, B.A.O. 1985. *Ethics and the limits of philosophy*. Cambridge, MA: Harvard University Press.
Wright, S. 1932. The roles of mutation, inbreeding, crossbreeding and selection in evolution. In *Proceedings of the sixth international congress on genetics*, ed. D.F. Jones, 2 vols., vol. 1, 356–366. New York: Brooklyn Botanic Garden.

Chapter 6
Towards EcoEvoEthics

Patrick Blandin

Abstract Ecology long considered the natural world as an "equilibrium world". This view culminated in the 1950s with the ecosystem paradigm, which was strengthened by the idea that the reciprocal selection of interacting species should produce ecological stability. At the end of the 1940s, Aldo Leopold's Land Ethic valued the stability of natural communities, and the balance of nature became a key issue for conservationists. Nowadays, there is a shift towards a co-change paradigm: interacting biological and non-biological entities are co-changing through a transactional web that forms the biosphere. Consequently, as ecology meets evolution, the conservation target must shift from the stability of ecological systems to their adaptability. Simultaneously, there is a need for an eco-evolutionary ethics which assumes that we and our co-evolving aliens are living in a changing world. Difficult issues should therefore be addressed, such as the uniqueness and intrinsic value of living entities versus the substitutability of functionally redundant species, and the evolutionary value of diversity. Finally, beyond the biocentrism versus anthropocentrism debate, this EcoEvoEthics should affirm that a thing is right when it tends to enhance the biosphere's capacity to evolve.

In 1949, Aldo Leopold justified his proposal for a new ethic – the Land Ethic – explaining that, in humanity's history, there has been an extension of ethics: ethics dealt first with the relation between individuals, and later with the relation between the individual and society. Leopold considered this extension as an evolutionary process, ethics being possibly a kind of community instinct providing guidance for meeting ecological situations. Stating that there was as yet no ethics dealing with people's relation to ecological communities, Leopold wrote that the extension

P. Blandin (✉)
Department Men-Natures-Societies, Ecoanthropology and Ethnobiology,
Muséum National d'Histoire Naturelle, 75005, Paris, France
e-mail: blandin@mnhn.fr

D. Bergandi (ed.), *The Structural Links between Ecology, Evolution and Ethics: The Virtuous Epistemic Circle*, Boston Studies in the Philosophy of Science 296,
DOI 10.1007/978-94-007-5067-8_6,

of ethics to this third element was an evolutionary possibility and an ecological necessity. This idea should be considered in the light of evolutionary ethics, which was initiated by Thomas Henry Huxley and developed as an important field during the very last decades, though not without debate (Wilson 1975; Ruse 1986, 2009; Ayala 2006; Boniolo and De Anna 2006). The fundamental idea is that the emergence of the human capacity to elaborate Ethics could be an evolutionary process, shaped by natural selection: ethical behaviour should enhance the probability of group survival. Leopold's "environmental ethics," which deals with principles of nature conservation, should appear as a significant step in humanity's search for survival within the fragile community of life on earth.

Fifty years after Leopold's Land Ethic, the Chilean philosopher Ricardo Rozzi (1999) proposed checking on the existence of reciprocal influences between ecosystems theory and environmental ethics as a possible illustration of his general statement: "Ethics and science establish a dialectic interrelationship that evolves historically through mutual and successive modifications." In this chapter, I shall explore the way ethics for nature conservation and ecology entered into such a dialectical process, and I shall argue that, as ecology undergoes a dramatic change of paradigm, from "equilibrium" to "change," environmental ethics must acquire a genuine evolutionary dimension.

6.1 An Equilibrium World and the Ecosystem Paradigm

Ecology, as a scientific field, has its roots in the nineteenth century (Golley 1993; Acot 1998). The structuring of natural communities gradually became a central topic with the development of descriptive studies for terrestrial and aquatic communities of species composition, of phenology, and of relations with environmental factors. Limnologists played an eminent role in this process. In 1887, the American Stephen Alfred Forbes, describing the lake as a "microcosm," focused on the concept of a "community of interest," defined as a community of interacting species shaped by natural selection, the "beneficent power" of which compelled "such adjustments of the rate of destruction and of multiplication of the various species as shall best promote this common interest" (Forbes 1887, 87). Clearly, Forbes considered that the selection process produced an equilibrium that was steadily maintained (barring dramatic changes in local conditions), which for all the parties involved achieved the greatest good permitted by the circumstances.

Plant ecologists have also become very active since the end of the nineteenth century. In 1910, the Frenchman Charles Flahaut and Carl Schröter from Switzerland coined the concept of a plant association: a plant community with a precise species composition, adapted to precise ecological conditions. The American Frederic Clements elaborated the theory of the development of plant communities towards an equilibrium state, the "climax" (Clements 1916). Ideas converged on both sides of the Atlantic: phytogeographers recognized that plant communities adapt to their environment and, once adapted, remain at equilibrium unless the ecological context changes. These ideas had obvious similarities with Forbes's conception. Thus, pioneer ecologists were giving primacy to the concept of natural equilibrium.

The English botanist Arthur George Tansley (1935) elaborated the ecosystem concept as a rebuttal to a burst of papers published by John Phillips, a South African botanist who applied Jan Christian Smuts's concept of holism to ecological communities and promoted Clements's metaphoric view of the plant community as a super-organism (see Bergandi 1999). Tansley claimed that the smallest unit of nature – the "ecosystem" – includes not only plants, but all the biotic components (plants, animals, microorganisms) and their physical environment (the abiotic context). Tansley considered that, within an ecosystem, biotic and abiotic factors are in a relatively stable dynamic equilibrium.

The limnologist Raymond Laurel Lindeman (1942), who worked with George Evelyn Hutchinson at Yale University, developed a brilliant synthesis between quantitative research on food webs, Clements's succession and Tansley's ecosystem. He introduced the concept of a "trophic equilibrium," a dynamic process of the use and regeneration of nutriments, supported by a continuous energy flow. In the spirit of Lindeman's paper, in 1953 the American ecologist Eugene Pleasants Odum, assisted by his brother Howard Thomas Odum, published *Fundamentals of Ecology*, the keystone book of modern ecology. Lindeman's and Odum's approaches were both systemic, with Odum's book giving the ecosystem concept a paradigmatic role (Bergandi 1995).

In the USA, ecology rapidly met system analysis and cybernetics, thanks to the availability of the first digital computers (Golley 1991). The use of ecosystems analysis and of cybernetic models, or "eco-cybernetics" (Bergandi 2000), has been at the heart of ecological research for decades. The ecosystem was considered a cybernetic entity, structured by interactions within species and the environment, and maintained in a dynamic equilibrium by feed-back processes. Moreover, with the development of the concept of a dynamic equilibrium sustained by a continuous energy flow, the thermodynamics of open systems consolidated ecosystem theory.

Darwin's theory was used early on to justify ecosystem stability. Forbes (1887) was perhaps the first to put forward the idea that species communities are shaped by natural selection, with their stability resulting from the tight adjustment of the various species dynamics. A century later, John A. Wiens (1984, 440) wrote: "Ecology has a long history of presuming that natural systems are orderly and equilibrial (the 'balance of nature' notion; . . .), and the infusion of evolutionary thinking into ecology strengthened this view, providing a mechanism (natural selection) that may lead to the development of optimally structured communities." In the meantime, Tansley (1935) supposed the existence of a kind of competition between ecosystems, with those reaching a more stable equilibrium having a longer survival time. In this way, evolutionary theory favoured a non-evolutionist ecological thought, stability being considered as the normal state of ecosystems, when undisturbed by humans. A very emblematic reflection of this conceptual framework can be found in *Evolutionary Ecology* (Shorroks 1984), a volume that resulted from the 23rd Symposium of the British Ecological Society.

The final chapter is entitled "Genetic diversity and ecological stability," in which the author concludes "that the plasticity produced by genetic diversity as a result of ecological interactions is an important factor in maintenance of persistence in ecosystems" (Mani 1984, 394).

For decades, ecologists worked mainly on situations existing at one particular moment of time, in one particular area. They could work with peace of mind: the "equilibrium competitive community paradigm" (Wiens 1984, 456), reinforcing the ecosystem paradigm, provided a perfect umbrella. As stated by Robert E. Ricklefs (1987, 167), "present-day ecological investigations are largely founded on the premise that local diversity – the number of species living in a small, ecologically homogeneous area – is the deterministic outcome of local processes within the biological community." This hypothesis gave hope that there are general laws connecting ecological context, competition, natural selection, and the species diversity of communities.

Paradoxically, while the synthetic theory of evolution was spreading among biologists, providing a general framework for biology, it supported the development of an ecological theory favouring a static view of nature: contemporary ecosystems were supposed to have reached a stable state, shaped in the past and now maintained by natural selection. In this intellectual context, man was necessarily viewed as an external, perturbing factor in otherwise "perfect" nature.

6.2 Protection of Nature: The Path to Ecology

During the eighteenth century, emblematic decisions were taken for the protection of forests in some tropical islands in French and British colonies. These decisions were inspired by "proto-ecological" conceptions that dealt with relationships between forests, climate, and water availability, and they provided models for conservation actions, for example in India (Grove 1992). But areas were also protected in the nineteenth century for purely aesthetic reasons. In France, as early as 1853, the Barbizon Painters, who worked in the Fontainebleau Forest, succeeded in obtaining the creation of a small (624 ha) "artistic reserve," where nature was protected for its landscape beauty. In the United States, the first national park, Yellowstone, was created in 1872 in an effort to protect vast areas of wilderness and satisfy public aesthetic and moral yearnings and a thirst for outdoor recreation. At the end of the century, the preservationist John Muir, influenced by Ralph Waldo Emerson's and Henry David Thoreau's philosophy of nature, was an active promoter of "wilderness" preservation as the purest representation of divine creation. This same period saw rising awareness of the extinction of wild species, with the most emblematic action being William Temple Hornaday's fight for the survival of the American bison (Hornaday 1889). At the international level, agreements were signed in 1883 in Paris for the protection of sea mammals in the Bering Sea and in 1902 for the preservation of "useful birds."

US President Theodore Roosevelt was influenced by John Muir, but much more by the forester Gifford Pinchot, who developed a utilitarian conception of nature conservation. Roosevelt planned an international conference on conservation to be held in The Hague in 1909, but his successor killed the project (Holdgate 1999). Europe also saw the rise of movements for the protection of nature. In 1905, the "Congrès International pour l'Art Public," held at Liège, adopted a resolution for the creation of natural parks, presented by a French agronomist and lawyer, Raoul de Clermont. An international movement was launched in 1910 by the Swiss naturalist Paul Sarasin, and an international office for the protection of nature was created in 1913, but this trend was interrupted by World War I. The movement regained momentum in 1923 with the First International Congress for the Protection of Nature, held at the Muséum National d'Histoire Naturelle, in Paris, with Raoul de Clermont as its general secretary. The congress considered many aspects of nature conservation, including fauna, flora, fossils and minerals deposits, natural monuments and landscapes.

Academic ecology was notable for its absence at the Paris congress. There was no consideration of communities, in Forbes's or Clements's sense. Zoologists and botanists focused on endangered species and habitats (sometimes using the word "station," meaning the local site of a plant association), but without a genuinely ecological approach. It was only 25 years later, in 1948, when the International Union for the Protection of Nature (IUPN)[1] was created by an international conference held in Fontainebleau, that the Union founders advocated the development of research in ecology (UIPN 1948). For the first time in an official context, conservationists recognized that ecology should be a key science for the protection of nature. The IUPN held its first Technical Conference at Lake Success (USA) in 1949 (UIPN 1950). A preparatory document was published (UIPN 1949), including parts of a previous paper by Jean-Paul Harroy, the IUPN General Secretary, who pointed out a radical change taking place in the protection of nature. Harroy argued that protection should no longer limit itself to the sentimental point of view that had persisted for so long in protectionist thinking. Especially as mankind became increasingly anxious due to the sombre predictions of economists, protection needed to take on board a utilitarian perspective. This change – with the idea that nature conservation and the economy are linked – pointed to the crucial need for a scientific study of natural communities, and ecology appeared to be the appropriate science.

As a matter of fact, an important portion of the Lake Success Conference was devoted to ecology. This was introduced by a paper entitled "Protection de la nature et écologie," presented by a French biologist, Georges Petit (1950). Petit, who had participated in the Paris congress in 1923, emphasized the fact that relationships between the protection of nature and ecology had been widely neglected, as the former had long been motivated only by aesthetic and moral concerns. Petit said that, by the way, protection was considered for a long time as no more than an art added to the study of nature.

Although the term "ecosystem" had been proposed more than 10 years earlier (Tansley 1935), it was never referred to at Lake Success. The "association" or "natural community" was the central concept. Petit (1950), for example, focused his paper on "vegetational complexes." He explained that, in tropical countries, the vegetation, not modified by man, was a complex that resulted from an evolutionary process, with its physiognomy and composition tightly linked to the present local conditions: a perfect image, he said, of what botanists had named the climax. In this kind of complex, Petit added, phytosociologists see the expression of a stable, equilibrated plant association: Flahaut's and Clements's influences were obvious. In fact, the concept of a "natural equilibrium," or a "balance of nature," was at the heart of the conference discussions, with the disruptive factor in this natural equilibrium ranging from the introduction of an exotic species to the extermination of big game herds and the unwise use of powerful modern insecticides, as explained by the IUPN General Secretary in his introduction to the proceedings (Harroy 1950).

It is important to grasp the conceptual situation at this moment when the conservationist movement met ecology. Harroy's viewpoint (1949) illustrates conservationists' expectations. He considered that, to efficiently protect useful natural associations, man must have studied them carefully beforehand. But to study these associations in the best conditions – Harroy wrote: "in the state of a pure body" – man must have protected them. In appropriate and sufficiently vast areas, shielded from human influences that mask and distort fundamental processes, researchers should attempt to observe these processes and to order them into laws. This statement is symptomatic of what we can call the "virgin nature ideology," which considers man as an external factor, whose interference makes it impossible to understand the real properties of nature.

6.3 Ecocentrism, the Ethical Counterpart of the Ecosystem Paradigm

The Paris Congress report is important for understanding the ideological background and scientific context of nature protection at the beginning of the twentieth century (Clermont et al. 1925). In his closing address, Professor Louis Mangin, the Director of the Muséum National d'Histoire Naturelle, expressed the idea that conservation was necessary not only for aesthetic or moral reasons, but also for practical ones: natural richness was being destroyed, when prudent use would allow its perpetuity. This view was rather similar to Pinchot's conception of natural resources conservation. Nevertheless, during the congress, the dominant values underlining concern for protection were the rarity of species, the beauty or scientific interest of animals and plants, and the artistic, historical or legendary interest of natural sites and landscapes. Protection ethics, at that time, reflected a mixture of biocentrism – awareness about endangered species that science would never be able to recreate, as Louis Mangin put it – and cultural anthropocentrism.

Introducing his Land Ethic, Leopold (1949, 214) wrote: "An ethic to supplement and guide the economic relation to land presupposes the existence of some mental image of land as a biotic mechanism. We can be ethical only in relation to something we can see, feel, understand, love, or otherwise have faith in." Leopold described a kind of virtuous circle, linking scientific knowledge and ethics, which could also be interpreted according to Rozzi's view of the interrelations between science and ethics (Rozzi 1999).

Leopold was not comfortable with the "balance of nature" image, despite its common use, probably because it provided no scientific view of reality. Thus, he proposed as a much truer image a concept employed in ecology, the "biotic pyramid," to evoke the complex web of food chains within a biotic community. He observed that the stability of this system proves it is a highly organized structure that functions through the co-operation and competition of its diverse parts. Moreover, Leopold considered that such stable, organized structures result from a long evolutionary succession of adjustments between parts: his vision was close to Forbes's conception of the community of interest, and to the views of Tansley, who considered that the degree of perfection of the ecosystem equilibrium is revealed by its level of stability (Tansley 1935).

Leopold did not use the ecosystem concept, which at that time remained uncommon, but it is obvious that his conception of "land" was fully congruent with this. The Land Ethic values both the integrity and the stability of the biotic community, and Leopold held that conservation is the effort to preserve the land's capacity for self-renewal. Clearly, in Leopold's mind, the preservation of the integrity of the land/ecosystem, that is the conservation of all its components, was the condition for maintaining its stability and capacity for self-renewal. Leopold's ethic was thus in tune with contemporary ecological knowledge, focusing on biotic communities and later on ecosystems: it has been characterized as "ecocentric," as it values interdependences between the diverse parts, including humans, of ecological systems (cf. Callicott 1989).

6.4 Ecology Meets Evolution: The Co-change Paradigm

As early as 1973, Amyan Macfadyen, in his presidential address to the British Ecological Society, noted that some ecologists argue that ecology, like human history, is concerned with unique events which are not supposed to be open to the "scientific method" (Macfadyen 1975). Recently, in the same spirit, Peter Taylor and Yrjö Haila (2001) have pointed out the on-going shift from an ecological theory that is willing to elaborate general laws towards theories that take into account historical contingency, non-equilibrium dynamics, and the uniqueness of many situations.

This conceptual shift was pinpointed in 1987 by Robert E. Ricklefs: ecologists, he said, were realizing that local diversity bears the imprint both of global processes such as dispersal and species production and of unique historical circumstances. Ricklefs emphasized the necessity to consider the balance both between local and regional processes and between short-term events and long-term processes in order to understand species diversity on a local scale. Considering that, through interactions between species, selection favours increased competitive ability and predator efficiency, he concluded that evolution, while fostering greater accommodation among coexisting species, ultimately tends to reduce species richness. Ricklefs affirmed that this reduction is balanced by the immigration of individuals from other areas, the variety of which depends both on regional processes such as the generation and dispersal of new species and on historical accidents and circumstances that are related to past climate history and the geographical position of dispersal barriers and corridors.

Moreover, Ricklefs underlined that the historical dimension of any ecological system results in a diversity of local situations. In doing so, he laid the foundations for the concept of the historical trajectory of an ecosystem: in its present state, any ecosystem is the product of processes that unfold over time, marking out a unique history. Interestingly, the term "historical ecology" was not coined by ecologists, but by anthropologists working on interactions between human populations and their ecosystems and landscapes. Key roles were played by Carol L. Crumley, with her studies on Burgundy (Crumley and Marquardt 1987) and her direction in 1994 of *Historical Ecology: Cultural Knowledge and Changing Landscape*, and by William Balée (1992, 1995), a specialist of Amazonia.

At the same time, landscape ecology also favoured an important conceptual shift, by introducing a new way of looking at the spatial organization and dynamics of ecosystems (see for example Forman and Godron 1986). Disturbance, which was previously considered as perturbing normal equilibria, now appeared as the driving process behind mosaic landscapes. This idea took shape progressively, with the book *The Ecology of Natural Disturbance and Patch Dynamics* (Pickett and White 1985) marking a milestone. Later, Wu and Loucks (1995) went so far as to describe the emergence of this field as a paradigm shift, from the "balance of nature" to "patch dynamics."

Nowadays, it is obvious that each ecosystem and each landscape is a step along a unique trajectory. As local, regional and global processes are continuously interacting, evolution can no longer be considered in the limited sense of species originations and extinctions: it is a global process of coevolutionary interactions and ecological changes.

The earth sciences have highlighted the permanent changingness of our planet and of life on it, over a history of increasingly intertwined relationships between biotic and abiotic processes. The evolution of the biosphere must therefore be considered as a web of interdependent trajectories. As part of broadening the perspective, the Israeli ecologist Zev Naveh (2000) suggested that the "Total Human Ecosystem" should be regarded as the highest coevolutionary ecological entity on earth: he considered evolution as a dynamic process of self-organization and coevolution in nature and human societies. I suggest calling the evolving ecological web the "transactional web" in the spirit of Dewey's transaction concept (Dewey and Bentley 1949; see also Bergandi and Blandin 1998), which was transposed to ecological systems by Hills (1974).[2] Transactions, i.e. simultaneous, reciprocally determined changes between interacting entities, occur between physical environments and living systems, as well as between coevolving species, on every scale through the transactional web. Clearly, a "co-change paradigm" is taking the place of the "equilibrium paradigm" (regarding the transactional framework see 1.2 by Bergandi in this volume).

6.5 An Eco-evolutionary Ethics Is Needed

Can the conservation of nature be a fight against change? Acknowledging the fact that nature is definitively not a subtle, integrated equilibrium, but is intrinsically chaotic, and thus unpredictable, the French philosopher Catherine Larrère (1997) asked whether the integrity and stability of the biotic community, emphasized by Aldo Leopold as key values, had any sense. If everything is changing, how can we know what is right, what is wrong? Ecocentrism was the ethics produced in interaction with the ecologists' conception of an "equilibrium world": it is ultimately insufficient to provide values and principles for action in an evolving world. The paradigm shift in ecology now calls for a new step in the development of environmental ethics (Blandin 2004).

In 1957, the first UNESCO director general, Julian Sorell Huxley, who played a fundamental role in the creation of the International Union for the Protection of Nature in 1948, published a book in which he stated that man has a responsibility for the whole future of evolution. The same idea was promoted later by Otto Frankel (1974), and then again by Otto Frankel and Michael Soulé (1981) in their seminal book, *Conservation and Evolution*. Considering "the more stringent requirements for long-term conservation, involving the maintenance of the evolutionary potential, the capacity to evolve in response to environmental change," Frankel and Soulé introduced the fundamental idea of "evolutionary potential." In this scientific context, the aim of nature conservation should be to preserve the biosphere's evolutionary potential, in order to maintain the sustainability of ecological processes, despite changes in the composition and organization of ecological systems. The biosphere's permanent adaptability becomes the target.

Hutchinson (1964), commenting on Forbes's microcosm, underlined the fact that Darwin and Forbes were conscious that the struggle for life produces harmony. Hutchinson considered it was possible to go further, because at any scale in the universe harmony implies diversity. As we lack a less diversified universe for comparison, Hutchinson said, we cannot know whether diversity is definitely a significant property of our Universe, but we feel that it could be important and we need to appreciate it properly. Ecologists had developed ideas on this point. Relationships between the stability of ecosystems and the diversity of their species have been explored at least from the end of the 1950s, and many ecologists supported the idea that the more diverse an ecosystem is, the more stable it is. In 1975, Daniel Goodman reviewed the empirical and theoretical attempts to check this idea of a direct relationship between the species diversity of a community and its stability. His conclusion was negative: at that time, the expectations of the diversity-stability hypothesis were not borne out by experiment, observation, or models. Nevertheless, Robert M. May (1984, 6–7) noted that, "The idea that complex ecosystems, with many species and a rich web of interactions, should be more stable than simple ones is an intuitively appealing one; it may seem that a community is better able to cope with disturbance if there are many alternative pathways along which energy and nutrients may flow." As a matter of fact, 20 years after Goodman's criticisms, the hypothesis was still being taken into consideration: Silver and her colleagues (1996), for example, argued that functional diversity, and not just species richness, is important in maintaining the integrity of nutrient and energy fluxes. However, these authors underlined that high species richness may increase ecosystem resiliency following disturbance, thanks to a high number of alternative pathways for the flow of resources.

Recently, an international group of ecologists reviewed the current state of knowledge, and concluded that experiments and models support the idea that ecosystem performance depends on species diversity (Hooper et al. 2005). Nevertheless, they focused more on relationships between diversity and "ecosystem services" than on the capacities of ecosystems to adapt and evolve. They forgot Hutchinson's path-breaking idea, expressed as follows: "Just as adaptive evolution by natural selection is less easy in a small population of a species than in a larger one, because the total pool of genetic variability is inevitably less, so it is probable that a group containing many diversified species will be able to seize new evolutionary opportunities more easily than an undiversified group" (Hutchinson 1959, 156). In an evolutionary perspective, the sustainability of an ecosystem implies not only functional continuity (which could result from alternation between redundant species), but also the persistence of its capacity to evolve, which depends on the ecosystem's levels of genetic and species diversities. I explored such ideas, proposing, schematically, two different adaptive strategies, called "cenotic" strategies (Blandin et al. 1976; Blandin 1980). On the one hand, the adaptability of ecosystems with a low species diversity would depend on the genetic diversity – and consequently on the adaptability – of a few species carrying out keystone functions. On the other hand, the adaptability of ecosystems with a high species diversity would depend on the existence of functionally redundant species with different ecological aptitudes, with some species substituting for others under new environmental conditions. These two ecosystem strategies were considered opposite poles of a gradual range of situations: the "evolutionary potential" of an ecosystem depends on a particular combination of species diversity and genetic diversity within each species. This combination results from the past trajectory of the ecosystem.

Nowadays, ecologists also recognize that, at the landscape scale, the diversity and spatial arrangement of ecological systems – their ecological diversity – influence their capacity to adapt, for example in a context of climate change. The diversity of living systems at any level of organization therefore appears not only as a condition for the short-term sustainability of ecosystems but also as an assurance of their adaptability and evolution over the long term. Consequently, an ecological and evolutionary ethic should give a landmark value to the diverse character of living systems, independently of any human-centred considerations. A fundamental question nevertheless remains, as was clearly expressed by Frankel and Soulé (1981, 7): "If as biologists we accept the proposition that life cannot continue without opportunities for evolution, there remains the question why we should be concerned about the continued existence of living organisms except on grounds of actual or potential use to our own species." This issue is topical for ethics: the long-term existence of the Biosphere could be a biocentric or an anthropocentric target.

Anthropocentric reasons for conserving biodiversity were stated in a very explicit manner in the foreword of the *Global Biodiversity Strategy*, published by international organizations to prepare for the Rio de Janeiro World Conference (Speth et al. 1992). The authors, heads of international organizations, expressed the official consensus prevailing at that time, which considered that the conservation of biodiversity is fundamental to the success of the development process, and that conserving biodiversity is not just a matter of protecting wildlife in nature reserves. They emphasized that conservation is also about safeguarding the natural systems of the earth, which are our life-support systems, as they purify the waters, recycle oxygen, carbon and other essential elements, maintain the fertility of the soil, provide food from the land, freshwaters and seas, yield medicines, and safeguard the genetic richness on which we depend in the ceaseless struggle to improve our crops and livestock.

These arguments are typical of a purely functional view that focuses on what is now called "ecosystem services"[3] (Millennium Ecosystem Assessment 2005), with no explicit evolutionary perspective. Actually, in a more or less detailed manner, and with more or less original examples, many conservationists put forward the present and future services that biodiversity ensures or will provide for humanity's benefit. But they don't answer the question posed by Frankel and Soulé: do we want the biosphere to continue because we value life for itself, or because we believe that the continuation of ecological processes is necessary for humanity's perpetuation in a changing world?

6.6 Uniqueness, Diversity, and Evolutionary Values

On 30 September 1948, at the opening session of the Fontainebleau conference where the IUPN was created, Julian Huxley, the UNESCO general director, evoked "the fascination of all these other manifestations of life which, though all products of the same process of evolution, yet are something in their own rights, are alien from us, give us new ideas of possibilities of life, can never be replaced if lost, nor substituted by products of human endeavour" (reported by Holdgate 1999, 32). In one admirable sentence, Huxley emphasized the diversity of life's manifestations, affirmed that they are alien from us, even if they (and we) are products of a unique history, and recognized that they have rights of their own, that they are definitively unique, and therefore cannot be replaced. In doing so, he revealed the complexity of the ethical issues that nature protectionists have to face.

The notion that all other living creatures – animals, plants, bacteria, and even viruses – being ultimately unique aliens, have their own rights to exist, as Huxley said, stands at the very heart of the debate in environmental ethics. Can we consider the extinction of Ediacara species, 540 millions years ago, from an ethical perspective? Probably not, as nobody was advocating their protection. Considering the eco-evo-dynamics of the biosphere, the death of the last panda – the emblematic endangered species for many conservationists – will have no more consequence for the continuation of the biosphere than the death of the last individual of the dinosaur species had. The transactional web will go on. Nevertheless, if humans have an interest in biosphere history, then each living being should have a value as part of the living memory of past evolution. Therefore, to consider that each living being, and the whole living community, have to be protected "no less than the heritage of our culture" (Ghilarov 2000) is a cultural choice that confers anthropocentric values on living entities. On the other hand, to affirm that any living being has an intrinsic value and warrants respect, as the unique result of a particular trajectory within the evolutionary process, or just because it is a contributor to the transactional web, independently of any human interest, is typical of a biocentric attitude. In some respect, Huxley's view prefigured biocentrism in its later incarnation (Taylor 1986; Rolston 1988). This ethic clearly inspires the first of the "Ten Principles for Conserving Biodiversity" stated in the *Global Biodiversity Strategy* (WRI, IUCN and UNEP 1992): "Every form of life is unique, and warrants respect from humanity."

To recognize that aliens have value independently of humans' interests could be considered as a noble, altruistic effort, a rebuttal to arrogance towards the natural world. But a radical biocentrism could legitimate a radical conservationism, justifying the expulsion of people from their territory to protect a supposedly virgin nature. Some philosophers disagree with such a radical biocentrism. For example, Bryan G. Norton (1988, 201) believes "that species have value as a moral resource to humans, as a chance for humans to form, re-form and improve their own value systems." I think that this statement, which evokes useful relationships between man and other species, bears some similarity to Huxley's suggestion that, being alien from us, other living species offer us different images of life, recalling that man is only one life form among many. This position is evidently anthropocentric, but reflects a wider perspective than the basically utilitarian standpoint, as Norton (1984) has underlined in a previous paper.

An evolutionary perspective cannot consider living beings as only results of the past. Let us express Darwin's fundamental principle as follows: because living beings are different, the adaptability of the systems they form becomes possible through the selection of those that are better adapted than others to new ecological contexts. Therefore, any being, because it differs in some manner from others, has a value for its contribution to adaptability, independently of human-centred considerations. Concrete differences between living beings within any system make this system "bio-diverse." When first formulated, "biodiversity" was just a neologism, a "passe-partout" useful for communication between people (who were not certain that they were speaking about the same thing). Now, even if "biodiversity" is too often considered as a kind of entity, and abusively substituted for nature, it is no more than a collective attribute of any assemblage of living entities that differ from each other. As such, biodiversity can be given a value. More precisely, if we understand that highly biodiverse systems are more adaptive than less biodiverse ones, we can recognize that biodiversity has an "evolutionary value."

Such an approach may have contradictory consequences: in order to avoid the loss of any living being that could contribute to the adaptability of the system it is part of, under circumstances we cannot predict, we should preclude any modification of the present biodiversity. In order to allow change, conservation will refuse any change: should an "evolutionary ethics" favour "fixist" conservation practices? Here, we are at the core of the "substitution problem," which has been brilliantly discussed by Dieter Birnbacher (2004). Let us consider Darwin's principle once again. Under new local conditions, better-adapted entities will continue to contribute to the transactional web, while others will no longer be able to participate and so will disappear. One current view in ecology is that functionally redundant species can exist within an ecosystem, with some species dominant under certain conditions while others replace the former when conditions change. Functional redundancy is a necessary condition for substitutability, and substitutability allows the continuity of ecological processes. This view supports the anthropocentric concept of "ecosystem services continuity" (see for example Millennium Ecosystem Assessment 2005). Nevertheless, some ecologists fight again the redundancy concept; Alexis Ghilarov (2000), for example, argues that each species definitely has a specific array of roles and that redundancy between species concerns only, at best, some roles that they effectively share.

In order to discuss this issue, Birnbacher (2004) opposes both economists and environmentalists: the former tend to consider substitutability to be the rule, the latter the exception. As a matter of fact, approaches to an economic valuation of biodiversity face a dilemma. In France, for example, a group of experts tried such a valuation for the government (Chevassus-au-Louis et al. 2009). They "solved" the dilemma by making a distinction between "exceptional biodiversity," i.e. rare species or ecosystems which cannot be valued (as historical, unique monuments) and "general biodiversity," composed of substitutable species, which produces ecosystem services. Interestingly, these species were valued not one by one (a highly difficult, if not impossible, task) but collectively, by measuring ecosystem services (i.e. timber production, carbon dioxide fixation, etc., by a hectare of an "ordinary" forest).

Consider first the problem of exceptional biodiversity. Rare species are highly valued by conservationists mainly because they are rare (a circular valuation!), perhaps under threat, or even at risk of extinction. In an evolutionary perspective, even if they have no importance for the transactional web, their memory significance should have a consensual value. Nevertheless, as is argued by Birnbacher (2004), historical values cover only a small part of what we value in nature. This is correct; for example, there are different ways of being rare. In many ecosystems, "keystone species," which generally are represented by only a few individuals, play dramatic roles, for example as regulators of various populations. Other species are numerically rare, and of secondary importance for ecosystem functioning, but may be important contributors to evolutionary potential, and perhaps will play essential roles in the future. Think of the discrete mammals "waiting for the extinction of the dinosaurs"... We can therefore value a rare species as a threatened part of life's memory, as a significant contributor to ecological processes, or as a future important player in the web of ecological transactions.

Now let us consider the species that contribute to "general biodiversity." Do species that can substitute for one another have the same value? Is this value linked to the role the species are able to fulfil? Is it linked to the capacity of a given species to substitute for another? In this case, the value of a particular species can change in accordance with the characteristics of the new context provoking the substitution: one context can favour the substitution of species S1 by species S2, while another will favour the substitution of S2 by S1... or by S3, etc. One possibility is to say that the value of a species depends on how many other species it is capable of replacing. If we follow this approach, we will give higher values to generalist species, even to invasive ones. Many conservationists will not agree.

Birnbacher (2004) concluded his philosophical tour through the "substitution problem" by considering relational values, when entities are objects of love, awe, admiration, or some other sentimental attachment. This is a very interesting point. Does that mean that no satisfactory solution can be found to give a rational foundation to the valuation of nature? At this point, I remember the conclusion of Jean Dorst, in his pioneering book *Avant que nature meure* (Dorst 1965). Dorst said that we have enough rational reasons to preserve nature, but nature actually will be preserved if we love it.

6.7 Conclusion

Let me close the circle. Hutchinson (1959, 157), at the end of his so important paper about diversity issues, raised a "metaphysical general point":

> The evolution of biological communities, though each species appears to fend for itself alone, produces integrated aggregates which increase in stability. There is nothing mysterious about this; it follows from mathematical theory and appears to be confirmed to some extent empirically. It is however a phenomenon which also finds analogies in other fields in which a more complex type of behavior, that we intuitively regard as higher, emerges as the result of the interaction of less complex types of behavior, that we call lower. The emergence of love as an antidote to aggression, as Lorenz pictures the process, or the development of cooperation from various forms of more or less inevitable group behavior ... are examples of this from the more complex types of biological systems.

These considerations are in harmony with Thomas Henry Huxley's philosophy, and they are congruent with Leopold's views on the evolution of ethics and also prefigure the hypotheses explored in depth by modern Evolutionary Ethics. Francisco Ayala (2006), for example, argued that the human potential to develop ethics has been shaped by biological evolution, but that our ethics are products of human history, including social and religious traditions. In this context, the construction of Environmental Ethics could be interpreted as a cultural, scientifically inspired process, enhancing humanity's adaptation to the Biosphere. This is obviously the position of Rozzi, who wrote (1999, 920):

> Instead, the interrelations between ecological-evolutionary sciences and environmental ethics can be understood as a dynamically and intimately bonded unit: under this unifying perspective, ecologists and eco-philosophers can overcome the schism between objective knowledge and subjective morality, recovering the link between theory and praxis, between the ways of knowing about nature and the ways of inhabiting the natural world.

Rozzi's reflections support the idea that environmental ethics evolves through a transactional process. Nowadays, the on-going substitution of the "equilibrium paradigm" by the "co-change paradigm" is producing an "eco-evo-ethics," based on the evidence that we live in a permanently changing world. This is a troubling idea: it undermines certainties, and promises no Eden to come. It obliges humans to understand that, while they try to build more stable, more comfortable environments, they always produce change, without being sure that it has any sense. At the same time, they know that they are becoming able to orientate change processes. Because humans are players in the transactional web, the Biocentrism vs. Anthropocentrism debate is obsolete. The aim today should be to organize conviviality with the Biosphere, to create optimal conditions for Man-Nature coevolution: we have to organize the transactional interplay. But can we refer to stable values? Values too are changing through science-ethics transactions. Therefore, perhaps only one eco-evo-ethics principle can be proposed. Written in Leopold's style, it will affirm: "A thing is right when it tends to enhance the Biosphere's capacity to evolve. It is wrong when it tends otherwise." As we occupy some place on Earth, we reduce space for our companion species. At any time, in any place, people will have to make choices. Science will provide tools; respect and love for our coevolving aliens will suggest guidelines.

Notes

1. In 1956, the name was changed to International Union for Conservation of Nature (IUCN).
2. According to Hills (1974), a transaction is a category of interactions which depend not only on the nature of the two elements apparently interacting, but on the nature of a majority (even the totality) of the elements interacting within the whole ecosystem, therefore considered as a "transactional totality." Bergandi (2007) proposed a more global approach and coined the "transactor" concept to characterize the co-changing "unit" formed by any supposed entity and its environment.
3. It is interesting to note that, at the 1948 Fontainebleau conference that created the IUPN, Julian Huxley used the expression "services écologiques officiels" in French in the conference report (UIPN 1949).

References

Acot, P. (ed). 1998. *The European origins of scientific ecology*, 2 vols. Amsterdam: Overseas Publishers Association/Gordon and Breach Publishers/Editions des Archives Contemporaines.

Ayala, F. 2006. Biology to ethics: An evolutionist's view of human nature. In *Evolutionary ethics and contemporary biology*, eds. G. Boniolo and G. De Anna, 141–158. Cambridge: Cambridge University Press.

Balée, W. 1992. People of the fallow: A historical ecology of foraging in lowland South America. In *Conservation of neotropical forests: Working from traditional resource use*, eds. K.H. Redford and C. Padoch, 35–57. New York/Oxford: Columbia University Press.
Balée, W. 1995. Historical ecology of Amazonia. In *Indigenous peoples and the future of Amazonia: An ecological anthropology of an endangered world*, ed. L.E. Sponsel, 97–110. Tucson: University of Arizona Press.
Bergandi, D. 1995. "Reductionist holism": An oxymoron or a philosophical chimaera of E.P. Odum's systems ecology? *Ludus Vitalis* 3(5): 145–180.
Bergandi, D. 1999. Les métamorphoses de l'organicisme en écologie: de la communauté végétale aux écosystèmes. *Revue d'histoire des sciences* 52(1): 5–31.
Bergandi, D. 2000. Eco-cybernetics: The ecology and cybernetics of missing emergences. *Kybernetes* 29: 928–942.
Bergandi, D. 2007. Niveaux d'organisation: évolution, écologie et transaction. In *Le tout et les parties dans les systèmes naturels*, ed. T. Martin, 47–55. Paris: Vuibert.
Bergandi, D., and P. Blandin. 1998. Holism vs. reductionism: Do ecosystem ecology and landscape ecology clarify the debate? *Acta Biotheoretica* 46: 185–206.
Birnbacher, D. 2004. Limits to sustainability in nature conservation. In *Philosophy and biodiversity*, eds. M. Oksanen and J. Pietarinen, 180–195. Cambridge: Cambridge University Press.
Blandin, P. 1980. Evolution des écosystèmes et stratégies cénotiques. In *Recherches d'écologie théorique. Les stratégies adaptatives*, eds. R. Barbault, P. Blandin, and J.A. Meyer, 221–235. Paris: Maloine.
Blandin, P. 2004. Biodiversity between science and ethics. In *Soil zoology for sustainable development in the 21st century*, eds. S.N. Shakir Hanna and W.Z.A. Mikhail, 3–35. Cairo: Cairo University.
Blandin, P., R. Barbault, and C. Lecordier. 1976. Réflexions sur la notion d'écosystème: le concept de stratégie cénotique. *Bulletin d'Ecologie* 7: 391–410.
Boniolo, G., and G. De Anna (eds.). 2006. *Evolutionary ethics and contemporary biology*. Cambridge: Cambridge University Press.
Callicott, J.B. 1989. *In defense of the land ethic: Essays in environmental philosophy*. Albany: State University of New York Press.
Chevassus-au-Louis, B., S. Bielsa, G. Martin, J.-L. Pujol, D. Richard, and J.-M. Salles. 2009. *Approche économique de la biodiversité et des services liés aux écosystèmes. Contribution à la décision publique. Rapport du groupe de travail*. Paris: Premier Ministre, Centre d'analyse stratégique.
Clements, F.E. 1916. *Plant succession: An analysis of the development of vegetation*. Washington, DC: Carnegie Institution of Washington.
Clermont, R. de, A. Chappellier, L. de Nussac, F. Le Cerf, and C. Valois (eds.). 1925. *Premier congrès international pour la protection de la nature, faune et flore, sites et monuments naturels (Paris, 31 May – 2 June 1923). Rapports, vœux, réalisations*. Paris: Société Nationale d'Acclimatation de France/Ligue Française pour la Protection des Oiseaux/Société pour la Protection des Paysages de France.
Crumley, C.L., and W.H. Marquardt (eds.). 1987. *Regional dynamics: Burgundian landscapes in historical perspective*. San Diego: Academic Press.
Dewey, J., and A.F. Bentley. 1949. *Knowing and the known*. Boston: Beacon.
Dorst, J. 1965. *Avant que nature meure*. Paris: Delachaux et Niestlé.
Flahault, C., and C. Schröter. 1910. Rapport sur la nomenclature phytogéographique. In *Actes du IIIème congrès international de botanique, Bruxelles 1910*, ed. E. de Wildeman, 131–142. Bruxelles: A. De Boeck.
Forbes, S.A. 1887/1925. The lake as a microcosm. *Bulletin of the Scientific Association of Peoria, Illinois* 1: 77–87. Reprinted in *Illinois Natural History Survey Bulletin* 15(9): 537–550.
Forman, R.T.T., and M. Godron. 1986. *Landscape ecology*. New York: Wiley.
Frankel, O.H. 1974. Genetic conservation: Our evolutionary responsibility. *Genetics* 78: 53–65.

Frankel, O.H., and M.E. Soulé. 1981. *Conservation and evolution*. Cambridge/New York: Cambridge University Press.
Ghilarov, A.M. 2000. Ecosystem functioning and intrinsic value of biodiversity. *Oikos* 90: 408–412.
Golley, F.B. 1991. The ecosystem concept: A search for order. *Ecological Research* 6: 129–138.
Golley, F.B. 1993. *A history of the ecosystem concept in ecology*. New Haven: Yale University Press.
Goodman, D. 1975. The theory of diversity-stability relationships in ecology. *The Quarterly Review of Biology* 50: 237–266.
Grove, R.H. 1992. Origins of western environmentalism. *Scientific American* 267: 42–47.
Harroy, J.-P. 1949. Définition de la protection de la nature. In *Documents préparatoires à la conférence technique internationale pour la protection de la nature, Août 1949, Etats-Unis*, ed. UIPN, 9–14. Paris: UNESCO.
Harroy, J.-P. 1950. Introduction. In *Conférence technique internationale pour la protection de la nature, Lake Success, 22–29 – VIII – 1949. Procès-verbaux et rapports*, ed. UIPN, vi–xi. Paris/Brussels: UNESCO.
Hills, G.A. 1974. A philosophical approach to landscape planning. *Landscape and Planning* 1: 339–371.
Holdgate, M.W. 1999. *The green web. A union for world conservation*. London: Earthscan Publications Ltd.
Hooper, D.U., F.S. Chapin III, J.J. Ewel, A. Hector, P. Inchausti, S. Lavorel, J.H. Lawton, D.M. Lodge, M. Loreau, S. Naeem, B. Schmid, H. Setälä, A.J. Symstad, J. Vandermeer, and D.A. Wardle. 2005. Effects of biodiversity on ecosystem functioning: A consensus of current knowledge. *Ecological Monographs* 75: 3–35.
Hornaday, W.T. 1889. The extermination of the American bison. In *Annual report of the board of regents of the Smithsonian Institution for the year 1887. Part 2*, 369–548. Washington, DC: Government Printing Office, Smithsonian Institution.
Hutchinson, G.E. 1959. Homage to Santa Rosalia or why are there so many kinds of animals? *The American Naturalist* 93: 145–159.
Hutchinson, G.E. 1964. The lacustrine microcosm reconsidered. *American Scientist* 52: 334–341.
Huxley, J.S. 1957. *New bottles for new wine: Essays*. New York: Harper.
Larrère, C. 1997. *Les philosophies de l'environnement*. Paris: Presses Universitaires de France.
Leopold, A. 1949. *A Sand County almanac, and sketches here and there*. New York: Oxford University Press.
Lindeman, R.L. 1942. The trophic-dynamic aspect of ecology. *Ecology* 23: 399–417.
Macfadyen, A. 1975. Some thoughts on the behaviour of ecologists. *Journal of Animal Ecology* 44: 351–363.
Mani, G.S. 1984. Genetic diversity and ecological stability. In *Evolutionary ecology. 23rd symposium of the British ecological society*, ed. B. Shorrocks, 363–396. Oxford: Blackwell Scientific Publications.
May, R.M. 1984. An overview: Real and apparent patterns in community structure. In *Ecological communities. Conceptual issues and the evidence*, eds. D.R. Strong Jr., D. Simberloff, L.G. Abele, and A.B. Thistle, 3–16. Princeton: Princeton University Press.
Millennium Ecosystem Assessment. 2005. *Ecosystems and human well-being: Synthesis*. Washington, DC: Island Press.
Naveh, Z. 2000. The total human ecosystem: Integrating ecology and economics. *BioScience* 50: 357–361.
Norton, B.G. 1984. Environmental ethics and weak anthropocentrism. *Environmental Ethics* 6: 131–148.
Norton, B.G. 1988. Commodity, amenity, and morality. The limits of quantification in valuing biodiversity. In *BioDiversity*, eds. E.O. Wilson and F.M. Peter, 200–205. Washington, DC: National Academy Press.
Odum, E.P. 1953. *Fundamentals of ecology*. Philadelphia: W.B. Saunders.

Petit, G. 1950. Protection de la nature et écologie. In *Conférence technique internationale pour la protection de la nature, Lake Success, 22–29 – VIII – 1949. Procès-verbaux et rapports*, ed. UIPN, 304–314. Paris/Brussels: UNESCO.

Pickett, S.T.A., and P.S. White (eds.). 1985. *The ecology of natural disturbance and patch dynamics*. San Diego: Academic Press.

Ricklefs, R.E. 1987. Community diversity: Relative roles of local and regional processes. *Science* 235: 167–171.

Rolston, H.III. 1988. *Environmental ethics? Duties to and values in the natural world.* Philadelphia: Temple University Press.

Rozzi, R. 1999. The reciprocal links between evolutionary-ecological sciences and environmental ethics. *BioScience* 49: 911–921.

Ruse, M. 1986. Evolutionary ethics: A phoenix arisen. *Zygon* 21: 95–112.

Ruse, M. (ed.). 2009. Introduction. In *Thomas Henry Huxley: Evolution & ethics*. Edited and with an introduction by M. Ruse, vii–xxxvi. Princeton/Oxford: Princeton University Press.

Shorroks, B. (ed.). 1984. *Evolutionary ecology. 23rd symposium of the British ecological society*. Oxford: Blackwell Scientific Publications.

Silver, W.L., S. Brown, and A.E. Lugo. 1996. Effects of changes in biodiversity on ecosystem function in tropical forests. *Conservation Biology* 10: 17–24.

Speth, J.G., M.W. Holdgate, and M.K. Tolba. 1992. Foreword. In *Global biodiversity strategy: Guidelines for action to save, study and use earth's biotic wealth sustainably and equitably*, eds. WRI, IUCN, and UNEP, v–vi. Washington, DC: World Resources Institute.

Tansley, A.G. 1935. The use and abuse of vegetational concepts and terms. *Ecology* 16: 284–307.

Taylor, P., and Y. Haila. 2001. Situatedness and problematic boundaries: Conceptualizing life's complex ecological context. *Biology and Philosophy* 16: 521–532.

Taylor, P.W. 1986. *Respect for nature: A theory of environmental ethics*. Princeton: Princeton University Press.

UIPN. 1948. *Union Internationale pour la Protection de la Nature, créée à Fontainebleau le 5 octobre 1948*. Brussels: UIPN.

UIPN. 1949. *Documents préparatoires à la conférence technique internationale pour la protection de la nature. Août 1949, Etats-Unis*. Paris: UNESCO.

UIPN. 1950. *Conférence technique internationale pour la protection de la nature, Lake Success, 22–29 – VIII – 1949. Procès-verbaux et rapports*. Paris: UNESCO.

Wiens, J.A. 1984. On Understanding a non-equilibrium world: Myth and reality in community patterns and processes. In *Ecological communities. Conceptual issues and the evidence*, eds. D.R. Strong Jr., D. Simberloff, L.G. Abele, and A.B. Thistle, 439–457. Princeton: Princeton University Press.

Wilson, E.O. 1975. *Sociobiology: The new synthesis*. Cambridge, MA: Belknap Press.

WRI, IUCN, and UNEP (eds.). 1992. *Global biodiversity strategy: Guidelines for action to save, study and use earth's biotic wealth sustainably and equitably*. Washington, DC: World Resources Institute.

Wu, J., and O.L. Loucks. 1995. From balance of nature to hierarchical patch dynamics: A paradigm shift in ecology. *The Quarterly Review of Biology* 70: 439–466.

Chapter 7
Ecology and Moral Ontology

J. Baird Callicott

Abstract The "superorganism" was the first paradigm in ecology, set out by Drude in Europe and Clements in North America. It was succeeded by the "ecosystem" paradigm, set out by Tansley, developed by Lindeman and consolidated by Odum, who, at the mid-point of the twentieth century, returned it to its superorganismic roots. The analogy of ecosystems to organisms could not withstand subsequent scientific scrutiny: ecosystems are too ill-bounded, porous, dynamic and artificial to be sufficiently like organisms to qualify as superorganisms. The reverse analogy – organisms to ecosystems – is more perfect. Humans and other organisms may be fruitfully conceived as superecosystems. One's very cells host mutualistic mitochondrial organelles; one's gut hosts a huge biodiversity of bacteria, as do the surface areas of one's body. In addition to the resident biota, abiotic materials (air, water, various nutrients) flow through oneself. This superecosystemic conception of oneself implies a relational – as opposed to a monadic – moral ontology. One's relationships – to other humans, to various kinds of animals, to one's various social and biotic communities, to the biosphere – generate a set of nuanced duties and obligations. One discharges such duties and obligations in a spirit of affection and pride, not in a spirit of begrudging self-sacrifice.

7.1 The Superorganism Paradigm in Ecology

Ecology emerged as a science during the final decade of the nineteenth century. The first paradigm in ecology was consolidated by Oscar Drude, in Europe, and Frederic Clements, in North America, during the first decade of the twentieth century. The putative objects of ecological study were plant "associations" or plant

J.B. Callicott (✉)
Department of Philosophy and Religion Studies, University of North Texas,
Denton, TX 76205, USA
e-mail: callicott@unt.edu

D. Bergandi (ed.), *The Structural Links between Ecology, Evolution and Ethics: The Virtuous Epistemic Circle*, Boston Studies in the Philosophy of Science 296,
DOI 10.1007/978-94-007-5067-8_7,

"communities," which were conceived to be "complex organisms" or, as they eventually came to be called, "superorganisms." Then it was also common in sociology – which emerged as an academic discipline at about the same time as ecology – to conceive of human societies and communities in holistic, organismic terms. So the ecological conception of plant associations or communities as superorganisms had a contemporaneous counterpart in the social sciences, especially in the French functionalism implicit in the work of Auguste Comte and explicit in that of Émile Durkheim. The biological and sociological aspects of the concept coalesced in the portrait of haplodiploidal insect societies (ants especially) as superorganisms by the German-educated Harvard entomologist William Morton Wheeler, a younger contemporary of Drude and an older contemporary of Clements. A good history of ecology is provided by Robert P. McIntosh (1985); a good history of sociology is provided by Randall Collins (1994).

The idea that ecology is the science of superorganisms – that the proper objects of ecological study are third-order organic wholes – seems, in retrospect, wildly metaphysical and hardly an auspicious beginning for a brand new science bidding for legitimacy in the panoply of established natural sciences like physics, chemistry, and biology. Some historians attribute the holistic ontology of early ecology to the influence of Drude on Clements and the influence of German idealism and romanticism – expressed in the works of Kant, Hegel, Goethe, and von Humboldt – on Drude.

While such historical influences may have played a role, I think that the inaugural superorganism paradigm in ecology is perfectly explicable without searching for origins in extra-scientific philosophy. The parent science of ecology is biology. Provided with a coherent and comprehensive organization by the publication of *Systema Naturae* by Carl Linnaeus in 1735, descriptive natural history provided a foundation for modern biology. Building on the Linnaean system of taxonomy, biology took its place among the modern sciences in the nineteenth century, spurred on by the improvements in microscopy early in the century, and provided a unifying theory by Charles Darwin in 1859, with the publication of *On the Origin of Species*. The principal objects of biological study were organisms. And the field of organismic biology was partitioned into taxonomy, anatomy, physiology, and ontogeny (or developmental biology). A good history of the emergence of biology as a natural science is provided by William Coleman (1977).

Clements (1905), whose influence among ecologists eventually eclipsed that of Drude, seems to have organized the new science of ecology analogously to the organization of its parent science, biology. He reasoned as follows. The first life forms were single-celled organisms. In the course of evolutionary time, symbiotic relationships among such organisms became so close that they merged to form higher-order multicelled organisms. Similarly, in the course of evolutionary time, the symbiotic relationships among multicelled organisms became so close that they merged to form still higher-order superorganisms. Until the invention of the microscope, we intelligent multicelled organisms were unaware of the existence of single-celled organisms. And until the advent of ecology, which provided the lens through which we could perceive them, we were unaware of the existence of superorganisms. As cells are to multicelled organisms, so multicelled organisms are to superorganisms. As organs are to multicelled organisms, so species populations are to superorganisms. Indeed, when ecology first emerged as a science, it was, by many, considered to be a branch of physiology. Just as intra-organismic physiologists proper studied the function of the various organs in multicelled organisms, so inter-organismic physiologists studied the functions of various species populations in superorganisms.

cells : multicelled organisms :: multicelled organisms : superorganisms

organs : multicelled organisms :: species populations : superorganisms

cells > organs > multicelled organisms > species populations > superorganisms

A hierarchy of biological organization as Clements imagined it to be: cells compose organs, organs compose multicelled organisms, organisms compose species populations, species populations compose superorganisms.

Ecology needed an eco-taxonomy, in addition to an eco-physiology in order for it to mirror its parent science, biology. Accordingly, Clements set out to provide such an eco-taxonomy by classifying the world's various superorganisms into types – arctic tundra, boreal coniferous forest, temperate mixed hardwood forest, sphagnum-tamarack, tupelo-cypress swamp, alpine meadow, prairie, steppe, etc., etc. Eco-anatomy then would be the study of the physical structure of the world's various superorganisms – for example, the oak-hickory forest floor, its understory, its canopy; the underground root system, the above-ground biomass, and the flowers and seed heads of a prairie; etc.; etc. Clements himself specialized in eco-ontogeny, the developmental sequence of a mature superorganism, which he termed "succession." After a ground-clearing natural disturbance like a wildfire or an anthropogenic disturbance like a clear-cut, "pioneer" species, mostly weedy annuals, would colonize the site, followed by brushy woody vegetation, followed by trees requiring full sunlight for germination and growth, followed by trees whose seeds and seedlings can grow in the shade of other trees. At such a point in the successional series, the superorganism would be fully mature and would reproduce itself until destroyed by an exogenous disturbance.

Biological/ ecological sub-fields	Organismal biology	Superorganismal ecology
Taxonomy	Genera (e.g. *Ursus*) Species (*arctos*, *maritimus*, etc.)	Genera (e.g., forests) Species (taiga, oak-hickory, etc.)
Anatomy	Structure of bones, muscles, nerves, etc.	Forest floor, understory, canopy, etc.
Physiology	Functional interrelations of organs	Functional interrelations of species populations

(continued)

Biological/ ecological sub-fields	Organismal biology	Superorganismal ecology
Ontogeny	Organismal development: larva→pupa→imago	Successional development: weeds→shrubs→light-loving trees→shade-tolerant climax

7.2 The Ecosystem Paradigm in Ecology

The superorganism paradigm in ecology, however, could not withstand closer, more systematic, and longer-term examination of the phenomena it was supposed to represent. Unlike multicelled organisms, superorganisms had no clear spatial boundaries, one putative type gradually merging and mixing with another. The posited developmental stages of superorganisms often failed to follow the expected sequence of stages similar to organisms that develop in a regular and determinate way – from say larva to pupa to imago. Nor had these putative "seres," as Clements called them, clear temporal boundaries – as one stage (or "sere") gradually merged into another. Nor did the whole successional process terminate, as Clements predicted, in a mature climax. Rather, change appeared to be unending, directionless, and stochastic. Paleo-ecologists finally delivered the *coup de grâce* to the superorganism paradigm in ecology. Upon examining pollens preserved in peat bogs and other anaerobic sediments, they discovered that past plant associations were constituted by different cohorts of species. The superorganism would have had to form, like Empedocles's man-faced ox and other similar phantasmagoria, with it organs (species populations) existing separately and then coming together in haphazard ways, existing thus for a time, then being driven apart (primarily by climate change), scattering away, and reassembling in novel combinations at a subsequent time.

The superorganism paradigm in ecology was eventually replaced by the ecosystem paradigm, which British ecologist Arthur Tansley sketched in his 1935 paper, "The Use and Abuse of Vegetational Concepts and Terms." By then, Charles Elton had made animal ecology a robust part of general ecology, and distinctive plant-animal assemblages were called "biotic communities." In other words, the plants *and* the animals living together and interacting with them and with one another constitute the biota of a place. Tansley's principal innovation was to add the abiotic materials cycling through the food chains of the trophic pyramid that Elton had so clearly described in his book, *Animal Ecology*. Inert elements – such as hydrogen, carbon, nitrogen, oxygen, phosphorous, potassium, calcium, and iron – are absorbed by the roots of plants from the soil and ultimately from the rocky substrate and combined with atmospheric carbon dioxide and water to assemble their living tissues. When herbivorous animals consume the tissues of plants, these elements in their new, complex chemical compounds pass into the bodies of animals. When those animals are, in turn, eaten by omnivores and carnivores, some of these materials pass into their bodies as well. Of course, all living organisms eventually die, whether consumed by others or not, and their bodies decompose with the help of saprophagous organisms. Thus these elements are returned to the soil; and from there, once again, they are incorporated into the living tissues of plants, thus completing a "nutrient cycle." Such processes of materials organization, metabolism, and decomposition formed, according to Tansley, "systems in the sense of the physicist." He dubbed them "ecosystems."

plants + animals = biota (à la Elton)
biota + abiotic materials = ecosystems (à la Tansley)

In 1942, in "The Trophic Dynamic Aspect of Ecology," Raymond Lindeman measured the radiant solar energy falling on the surface of a small pond – Cedar Bog Lake in the American state of Minnesota – and how much of it was converted by the autotrophs or primary producers (the green plants) of the lacustrine ecosystem into potential chemical energy available to the phytotrophes (the herbivores) that ate them. And so for each successive level of consumers in the pond and its many tangled food chains, Lindeman measured and tracked the flow of energy up to the large carnivores at the apex of the food web and finally down to the saprophagous organisms (the fungal and ultimately bacterial decomposers) in the pond's mud that garnered the residual chemical energy in the organic matter remaining in the dead tissues of plants and animals. In so doing, the decomposers made the elemental nutrients available for the autotrophs once again to synthesize into complex organic molecules, using the steady incidence of radiant solar energy to power the process of photosynthesis. Ecosystem ecology thus basically divided itself into two new sciences: (i) biogeochemistry, the quantitative study of the cyclical movement of elemental matter through ecosystems: and (ii) bio-thermodynamics, the quantitative study of the one-way flow of energy through ecosystems from solar source to ambient sink.

7.3 The Rise and Fall of Ecosystems as Superorganisms

The departure of Tansley from the superorganism paradigm of Clements has been greatly exaggerated by some historians of ecology, especially by Donald Worster (1977, 1985, 1994). To be sure, Tansley denied that ecosystems were literally big, spatially diffuse *organisms*, but he repeatedly characterized them as "quasi-organisms," and claimed that they existed in states of "dynamic equilibrium" and had evolved that highly organized condition of dynamic equilibrium by means of natural selection. In an influential paper published in 1969, "The Strategy of Ecosystem Development," the dean of mid-twentieth-century Anglo-American ecology, Eugene Odum, returned ecosystem ecology to its Clementsian organismic roots. Odum claimed, just as did Clements half a century prior, that "ecological succession involves the development of ecosystems; it has many parallels in the developmental biology [ontogeny] of organisms, and also in the development of human society." As Clements before him, Odum claimed that succession was determinate, directional, and predictable. The "mature" ecosystem attains the power to "control" its environment, by modulating its microclimate; by "closing" or "tightening" its biogeochemical cycling, thus arresting the loss of nutrients; by regulating the species populations of the organisms internal to it through competition and predation; by balancing production of organic matter or biomass with respiration, thus stabilizing its biomass; and by reducing entropy and increasing information (whatever that may mean). According to Odum, "The intriguing question is, Do mature ecosystems age, as organisms do? In other words, after a long period of relative stability or 'adulthood' do ecosystems again develop unbalanced metabolism and become more vulnerable to diseases and other perturbations?" (1969, 266).

Odum's (1969) representation of ecosystems is more the product of a vivid scientific imagination and wistful thinking than of sound empirical science. Succession is not, as a matter of fact, determinate, directional, and predictable. Ecosystems are not closed, self-regulating, and equilibrial. Yes, plant roots stabilize soils, retain moisture, slow nutrient loss, and plants modulate the microclimate, but ecosystems as such do not "control" their environments in any comprehensive way. The species populations within ecosystems are not balanced and constant; rather they oscillate wildly and are subject to frequent immigration, emigration, and extirpation. No more than the plant associations of Clements, the Tansleyan-Lindemanian-Odumesque ecosystems are hardly superorganisms. Further, by the time the ecosystem idea had been introduced into ecology by Tansley and completed by Lindeman, the Modern Synthesis of Mendelian genetics with Darwinian evolution-by-natural-selection had occurred in evolutionary biology. Unlike organisms, ecosystems have no genes for natural selection to sort, and thus they could not have emerged by means of the same evolutionary process, then widely accepted in evolutionary biology, by means of which organisms evolved. A good account of the current understanding of ecosystems is provided by Steward Pickett and Richard Ostfeld (1995); the putative consequences of the Modern Synthesis for natural selection operating beyond the organismic level of biological organization are set out by George C. Williams (1966).

Not only were the boundaries of ecosystems fuzzy and indistinct, they shift depending on the way ecosystems are interrogated by ecologists. In the first half of "The Trophic-Dynamic Aspect of Ecology," Lindeman essayed to measure and track the annual flow of energy through the Cedar-Bog-Lake ecosystem, the boundaries of which were thus marked by the pond's shoreline. In the second half of that article he essayed to account for its centuries-long succession from oligotrophic lake to eutrophic pond, eventually to become first a tamarack-sphagnum bog and finally a grassy moist meadow. The boundaries of the same Cedar-Bog-Lake ecosystem – which Lindeman was interrogating regarding the process of ecological succession – were thereby constituted by the surrounding watershed, from which nutrients and sediments washed into the pond, stimulated plant growth, and gradually filled it in. The very ontology of ecosystems thus seems shifting and elusive, driven by the vagaries of ecological inquiry. Ecosystems therefore appear to be artifacts of the methods of the science investigating them, not robustly existing independent entities (Allen and Starr 1982).

7.4 Organisms as Superecosystems

The stage is now set for a reverse metaphor. Ecosystems are not superorganisms. Rather, *organisms are superecosystems*. I first stumbled across this idea after a lecture I gave at the University of Maryland's Chesapeake Biological Laboratory in 1998. In response to my discussion of the superorganism paradigm in ecology the distinguished American ecologist Robert Ulanowicz agreed that ecosystems are not superorganisms; rather, he said, organisms were superecosystems. I replied that Ulanowicz was perhaps unwittingly trading on an ambiguity of the prefix "super." It was only later that I realized that he had made a profound and deeply significant observation. So, before explaining why the organism-as-superecosystem metaphor is both plausible and fruitful, let me note that ambiguity. In the ecosystem-as-superorganism metaphor, the prefix "super" means both (i) large – as in *super*market or as when you might go into McDonald's and say "*super*size me" – and (ii) of a hierarchically superior order of biological organization, a transorganismic order of biological organization. In the organism-as-superecosystem metaphor, the prefix "super" means superior in kind, not superior in size (bigger), nor superior in a conceptual hierarchy, as a genus is superior to a species or as a family is superior to a genus. In the organism-as-superecosystem metaphor, the sense of "super" is similar to that in *Super*man, the comic-book hero. Superman is a man, but a man superior to others in the manly physical virtues of strength, locomotive speed, jumping ability, and, indeed, in the manly moral virtues of justice, courage, chivalry, modesty, and general benevolence. Organisms may be plausibly and fruitfully considered to be superecosystems in the sense that organisms exhibit in a particularly superior way both the actual and imagined characteristics (virtues) classically attributed to ecosystems.

As to actual characteristics attributed to ecosystems, their metabolic processes and functions are carried out by organisms that are members of a variety of species populations – carried out, that is, by a diverse complement of plants and animals. Now, consider one's own organismic "self." Each one of one's own billions of cells is inhabited by other organelles called mitochondria, with their own DNA and enclosing membranes – and thus their own organismic identities and phylogenies. They are symbionts – more precisely mutualists, not parasites-and they supply our cells with adenosine triphosphate, the source of our cellular energy; and they also provide many other functional biochemical services. The endogenous human gut microbial community is so diverse and so well organized that it may be well said to be a smaller, internal superecosystem within the larger superecosystem that is a human organism (Zimmer 2011). And this too is typical of the conventional ecosystems that are the objects of ecological study – *macroecosystems* as we may now begin to call them. They too are hierarchically organized; that is, smaller ecosystems are nested within larger ones. For example, small geyser pools in Yellowstone National Park are ecosystems comprised of thermophilic bacteria. They are located in the much larger Greater Yellowstone Ecosystem – defined by the home ranges of its largest mammals (grizzly bears, wolves, bison, and elk) – which spills beyond the national park's political boundaries. A good account of the relationship between human cells and their mitochondrial symbionts is provided by Alberts et al. (2002); a good account of human intestinal ecology is provided by Marchesi and Shanahan (2007); a good account of hierarchy theory in ecology is provided by Allen and Starr (1982).

The human organism as superecosystem is similarly hierarchical. It is comprised of many other subsystems, outstanding among them the aforementioned intestinal superecosystems composed of a bewildering biodiversity of bacteria – up to 250 known philotypes, yet unresolved into narrower Linnaean taxa. And the sheer number of the population of individual bacteria residing in the roughly 10-meter-long human gut exceeds the number of individual human cells in the whole human organism by an order of magnitude (Sears 2005). In a healthy human superecosystem, many of the resident intestinal microbes are mutualists aiding in the digestion of food, while others are commensals, and still others parasites. And the whole microbial community is constantly resisting invasive pathogens – just as a healthy macroecosystem is believed by ecologists to do.

Healthy human skin is colonized by bacteria belonging to 19 different phyla, 205 genera, and some 1,000 species (Grice et al. 2009). The US National Institutes of Health has recently created a new Human Microbiome project (Proctor 2011). In a recent report (Grice et al. 2009) the authors expressly invoke the organism-as-superecosystem metaphor: "The skin is ... an ecosystem, harboring microbial communities that live in a range of physiologically and topographically distinct niches," the study authors write. "For example, hairy, moist underarms lie a short distance from smooth, dry forearms, but these two niches are as ecologically dissimilar as rainforests are to deserts." The human superecosystem is comprised of multicelled as well as single-celled organisms. Our eyebrows and eyelashes, for example, are the habitats to two species of parasitic mites, among the smallest of known arthropods, invisible to the naked eye (Rufli and Mumcuoglu 1981).

As to imaginary characteristics of macroecosystems, organismic superecosystems are homeostatic and self-regulating, maintaining a constant internal body temperature, a relatively narrow blood pressure gradient, a constant abundance and balance of electrolytes in the blood, a constant blood pH and salinity, a constant resting heart rate that elevates with exercise within a relatively narrow gradient, and so on (Buchman 2002). Organismic superecosystems resist and repel invasive exotic organisms that attempt to establish populations in them. They are spatially well-defined with clear boundaries. They develop in a determinate and predictable way. While most species of superecosystems do not control their environments, they resist the tendency of inorganic material structures toward entropic equilibrium with their environments. They exhibit low entropy and high information content.

Despite their ontological ambiguity, macroecosystems remain among the objects of investigation by ecologists. They are known to be open, not just to radiant solar energy and rainwater, but to the ingress and egress of various motile organisms and various aerosols and chemical pollutants. They are regularly buffeted by various natural and anthropogenic disturbances – by fire, wind, flood, drought, ice, pestilence, and disease – sometimes rebounding and recovering from these perturbations, sometimes being driven to alternate ecological phase states. That is, sometimes they "flip" into another type of ecosystem altogether. Macroecosystems are regulated not only by the climates in which they are located, but also by many other processes external to themselves as well as by processes, such as predator-prey dynamics, internal to themselves. For example, the El Niño/La Niña oscillation in the Pacific Ocean affects the drought and monsoon cycles of the American Southwest; and soil blown off of parched African fields is carried by winds to South America, thus supplying precious nutrients to the nutrient-poor tropical soils of the Amazon.

We organismic superecosystems are similarly open. We breathe in the atmosphere momentarily; we daily ingest water and metabolize the bodies of other organisms as food; would-be pathogens, parasites, and commensal microorganisms continually enter and leave our bodies. Put biogeochemically, the larger natural environment is flowing through us all the time. Put thermodynamically, we are dissipative systems, slowly burning the potential chemical energy stored in the organisms that we consume, thus to maintain a highly and hierarchically organized complex structure of elemental materials – hydrogen, carbon, nitrogen, oxygen, phosphorus, potassium, calcium, iron, and many others. Put metaphysically, we are as vortices in a flow of materials structured by the radiant energy of the sun; our identity as individual physical entities is as substantively indistinguishable from the physical environment as a whirlpool or standing wave is substantively indistinguishable from the water flowing through it. The individual identity of a watery vortex or standing wave is a matter of organizational structure, not of substance, and therefore a matter of internal, not external relationship to its environmental matrix.

7.5 Classical and Recent Expressions of the Organism as Superecosystem Concept

The conception of organisms – including, especially, human organisms – as superecosystems has had a fairly long run in environmental philosophy. It was expressed a quarter century or so ago by the late Norwegian philosopher Arne Naess (1979) as "Self-realization." Naess uses a capital "S" to distinguish his metaphysical notion of Self-realization from the popular idea of narcissistic self-realization repeated ad nauseam in pop psychology. A true understanding of oneself involves the continuity of oneself with one's environment, according to Naess. Thus for Naess, environmental ethics is less a matter of respecting the environmental Other than of *enlightened* Self-interest. A decade or so before Naess, the American human ecologist Paul Shepard expressed the notion of the self as a superecosystem in terms of a pond metaphor, almost surely alluding – whether deliberately or not – to Lindeman's sketch of a macroecosystem in his field-defining paper on the ecology of Cedar Bog Lake. According to Shepard (1969, 260):

> Ecological thinking ... requires a vision across boundaries. The epidermis of the skin is ecologically like a pond surface or a forest soil, not a shell so much as a delicate interpenetration. It reveals the self ennobled and extended, rather than threatened, as part of the landscape and the ecosystem, because the beauty and complexity of nature are continuous with ourselves.

Alan Watts, a British-born practitioner and popularizer of Zen Buddhism in the United States, expressed the organism-as-superecosystem idea a decade or so before Shepard – now about a half century ago:

> Theoretically, many scientists know that the individual is not a skin encapsulated ego but an organism-in-environment field. The organism itself is "focused" so that each individual is a unique expression of the behavior of the whole field . . . [But] there is a colossal disparity in the way in which most individuals experience their own existence, and the way in which the individual is described in such sciences as biology, ecology, and physiology. The nub of the difference is this: the way the individual is described in these sciences is not as a freely moving entity within an environment, but as a process of behavior which is the environment (Watts 1963, 55).

The conception of organisms – including, especially, human organisms – as superecosystems has been reprised by *à la page* social scientists and post-humanistic humanists who are developing a "vital materialism" and a "political ecology of things." Among the first to do so, Bruno Latour (1999), author of *Politiques de la nature*, is cognizant of the development of Deep Ecology, eco-philosophy, and environmental ethics among Anglophone philosophers, but digs deeper to challenge the residual modernist dichotomies that they have only critically glossed – such as the Nature/society dichotomy, the fact/value dichotomy, the subject/object dichotomy, and the primary-quality/secondary-quality dichotomy.

Jane Bennett, author of the much acclaimed *Vibrant Matter: A Political Ecology of Things*, is an American political scientist who writes as one recently woken from her dogmatic slumber – by Latour, among other "continental philosophers." Her erstwhile dogmatic slumber was imbued with dreams of Cartesian ratio-centrism, humanism, and the rightful human mastery of Nature. Her book illustrates and celebrates the "agency of matter" and the "force of things." Bennett (2010, ix) now recognizes that "the image of dead or thoroughly instrumentalized matter feeds human hubris and our earth-destroying fantasies of conquest and consumption. It does so by preventing us from detecting (seeing, hearing, tasting, smelling, feeling) a fuller range of nonhuman powers circulating around and within human bodies . . .. The figure of an intrinsically inanimate matter may be one of the impediments to the emergence of more materially sustainable modes of production and consumption." As a relatively recent migrant from humanism to "post-humanism," Bennett seems to be unaware of the explorations of Watts, Shepard, Naess, and other environmental philosophers who blazed the trail that she is now widening.

While fully informed by "theory," as contemporary "continental philosophy" is known among scholars of literary criticism and culture studies, Stacy Alaimo, an ecocritic, richly illustrates the idea of "trans-corporeality" as it is concretely and personally experienced in the lives of human superecosystems – especially in women's lives and as expressed in women's performance art. "By emphasizing the movement across bodies, trans-corporeality reveals the interchanges and interconnections between various bodily natures," writes Alaimo (2010, 2). "But by underscoring that *trans* indicates movement across different sites, trans-corporeality also opens up a mobile space that acknowledges the often unpredicted and unwanted actions of human bodies, nonhuman creatures, ecological systems, chemical agents, and other actors." Alaimo's work, synthesized in *Bodily Natures: Science, Environment, and the Material Self*, explores, as do the twenty-first century works of Latour and Bennett, the political implications of a conception of human organisms as superecosystems – although, of course, not by that name, which is Ulanowicz's casual coinage. The work of these authors indicates how the organism-as-superecosystem concept, by whatever name, is a point of convergence for a wide variety of communities of discourse – sociology (Latour), political science (Bennett), literary criticism (Alaimo) in the twenty-first century and environmental philosophy (Naess and others) in the twentieth century.

7.6 From a Modern to a Post-modern Moral Ontology

I turn now to the implications of the organism-as-superecosystem concept for moral ontology. In all these thinkers – from Watts and Shepard in the 1960s to Latour, Bennett, and Alaimo in the second decade of the twenty-first century – the Cartesian subject is the foil for a more ecological and vitally material sense of self and individuality. Watts (1963) characterizes the Cartesian subject as "a skin encapsulated ego," while Shepard (1969) characterizes the "self" somewhat more materially as "an arrangements of organs, feelings, and thoughts – a 'me' – surrounded by a hard body boundary: skin, clothes, and insular habits." And for him "the alternative is a self as a center of organization, constantly drawing on and influencing the surroundings, whose skin and behavior are soft zones contacting the world instead of excluding it." More abstractly expressed, oneself and other persons (which certainly would not exclude other animals) are nodes or nexuses in a skein of relationships – relationships with organisms both internal and external to one's superecosystem. Through one's superecosystem circulate water, various materials (both nutritious and poisonous) and the biogenic air. The material world, both in the form of inert matter and living matter – to invoke a distinction drawn by Vladimir Vernadsky (1929) – crosses the fuzzy and penetrable boundaries of the superecosystem that is oneself. Through the pores of one's skin, on the air one breathes into one's lungs, in the water one drinks, and in the food one eats.

This skein of relationships – the node or nexus of which is oneself – most importantly, for moral ontology, includes one's socio-cultural and economic relationships, as well as one's relationships with other-than-human organisms, and abiotic materials. One is the son or daughter of particular parents, possibly the brother or sister of particular siblings, possibly the parent of a son or daughter oneself. One is born in a particular country and learns to speak a particular language, absorbs a particular culture, and possibly learns to practice a particular religion. One is educated well or poorly, extensively or rudimentarily. One pursues a line of work and possibly embarks upon a life-long career or moves from job to job and possibly from place to place. After all such material, social, cultural, geographic, and economic relationships are catalogued in all their detail and nuance – something only to be conceived of as an ideal limit to an approximation – there is nothing left over. There is no Cartesian thinking thing, no Kantian transcendental ego, no Pythagorean-Platonic psyche, no Christian soul. That's it. That just is oneself, considered as a biogeochemical/ sociocultural/ economic superecosystem. Consciousness itself – the brute fact that the ego/ psyche/ soul was supposed to account for – is itself an epiphenomenon, an emergent property of the relationships among the neurological components of the sub-superecosystem, called the nervous system.

Remembering that Descartes is often called "the father of modern philosophy," modern moral ontology is based on a doctrine of external relations among moral monads – the psychological analogue to externally related material atoms. It is also based on an implicit theory of moral essences and accidents. Each such thinking thing – each ghost in its machine – is what it is independently of its relationships to other thinking things, from which it is absolutely separated by its own bodily cladding and that of the other thinking things. Notoriously, for Descartes (and conveniently for his follower, Malebranche) only the human body was inhabited by a thinking thing – a soul – while all other animals were mechanical automata, divinely crafted uninspirited machines. Thus they could be treated with moral indifference, just as one might treat a humanly crafted machine. While not going to such Cartesian-Malebranchean extremes, Kant made reason the essential attribute for the desert of moral regard, implicitly granting that, while other animals might be conscious, lacking reason they failed to qualify for moral regard or ethical consideration. Kant's contemporary Bentham took the bait and argued that the capacity for sentience was the essential capacity qualifying a being for moral treatment. And so an essence-accident template for moral ontology was established by both Kant and Bentham. One or another psychological essence – reason, sentience, conativity, etc. – qualified a being for moral consideration or ethical regard, and all other characteristics – gender, ethnicity, race, species – were morally irrelevant accidents. All beings possessing the moral essence are entitled to equal consideration of their equal interests by all moral agents. A good discussion of modern moral ontology is provided by Bernard Williams (1981).

Relationships, no less than gender, ethnicity, race, and species, are also moral accidents. A moral agent should be strictly impartial and give equal consideration to the equal interests of all moral patients, the class of which is defined exclusively by the moral essence. The prevailing essence-accident moral ontology of ethical theory from the late eighteenth through the late twentieth-century has become strained to the breaking point in the globalized world of the twenty-first century. More than seven billion people are presently affected, to one degree or another, by one's every choice through the global supply chain – purchasing an I-phone affects factory works in China, eating shrimp contributes to the destruction of a mangrove shoreline in Vietnam. Everyone everywhere is affected by simply turning on a lamp or driving a car – which contributes, however minisculey, to global climate change. Thus one may be partially responsible for flooding the atoll home of a Maldivian or Micronesian. And the mention of global climate change brings to mind the fact that one's present choices will affect not only spatially distant people and animals but temporally distant future people and animals as well – billions and billions of them – to all of whom one owes equal consideration of their equal interests. Or so requires the prevailing modern monadic moral ontology. The thought is mind numbing. Either we become morally overwhelmed and incapacitated or we re-envision moral ontology. A good discussion of the normative implications of the modern essence-accident moral ontology is provided by Catherine Wilson (2004).

7.7 Post-modern Ecological Moral Ontology: Toward an Erotic Ethic

Reflect instead on your own actual superecosystemic ethical practice. You will find that your social and environmental relationships generate the duties and obligations that you actually feel and to which you actually respond – with pleasure. The modern essence-accident moral ontology generates a one-size-fits-all ethics. The post-modern superecological moral ontology generates a finely nuanced hierarchical system of ethics. Consider familial duties and obligations to parents and children. One owes the former respect, deference, care in their old age; one owes the latter love, attention, material and financial support until they reach an age of independence, and perhaps a college education to boot. One discharges such duties with pride and pleasure, not begrudgingly. Further, one is certainly not under the same obligations nor has one the same duties toward other elders and other children. But one may have different duties and obligations to other elders and other children, depending on one's relationships with them. To elder neighbors, one should look in on them from time to time and run errands for them if they are sick or incapacitated. To elder fellow-citizens of our nation states, collectively, we owe a public pension, paid for by taxation, to pay for life's basic needs – shelter, food, medicine. One is obligated temporarily to take in and look after neighbor children when their parents are stranded at work or have an automobile accident and must spend the night in a hospital. To the children of our fellow citizens collectively we owe a public elementary and secondary education, also paid for by taxation. One has duties and obligations to one's colleagues that are different from those one has to family members and to unrelated neighbors – perchance to teach a class for them when they are off to a conference or sick in bed, maybe to read or comment on their scholarly writing.

One has duties and obligations to other animals that are similarly nuanced. Household pets are like second-class family members. One is obligated to feed them, to show them affection, to provide them with veterinary care when they are sick or injured. Within the constraints of pet ethics, one may, however, euthanize them when they are so old and infirm as to be a burden to themselves as well as to oneself. One has no such duties to farm animals, unless one is a farmer. But if one is a farmer, those duties are not the same as to one's household pets or to one's non-human fellow-workers – the sheep dog, the barn cat, the draft horse. Among the duties farmers have to farm animals is to feed them, shelter them, protect them from predators, and to slaughter them humanely and painlessly, if they are raised for the table or the market. One has duties to proximate wild animals, such as songbirds – among such duties is to prevent one's household pets from harassing or killing them.

As a citizen of a municipality and of a nation-state one has duties and obligations to our fellow citizens not owed to non-citizens. One has duties to our countries themselves, not reducible to those one owes one's fellow citizens – among them to serve in the armed forces if called to do so or in such alternative national services as the Peace Corps. And one has duties and obligations to one's fellow denizens of the global village, among them to oppose tyranny, to refuse to be complicit in human trafficking and other forms of egregious exploitation; and one has a duty to demand that one's own national government support international efforts to achieve economic and environmental justice for all. If Aldo Leopold (1949) is to be believed, one has duties and obligations to the fellow-citizens of our biotic communities *and* to those communities as such.

The threat of global climate change has become the greatest moral challenge of the twenty-first century. The modern monadic moral ontology implies a dreary zero-sum ethic of self-sacrifice for the sake of each and every one of the nameless and faceless global billions *and* to the indeterminate as well as the anonymous billions more to come in the coming centuries and millennia. To do one's bit to mitigate global climate change, one is called upon to curtail one's consumption so radically as to adopt a lifestyle that would make that of a cloistered Medieval monk seem voluptuous – or so argue such contemporary climate ethicists as Peter Singer, Henry Shue, Stephen Gardiner, and others committed to the modern monadic, essence-accident moral ontology. Equal interests must be treated equally, from this point of view, and thus to privilege one's own interests, in choosing among courses of action, is to be guilty of what Gardiner calls "moral corruption." Modern moralists hope to shame themselves and everyone else into attempting to give equal consideration, in choosing a course of action, to everyone's equal interests, even in a hyper-connected global economy and civilization. That hardly seems a path to achieving an ethical outcome; rather it has proved to be a path to ethical paralysis.

In sharp contrast, a post-modern moral ontology implies an inclusive ethic of care and concern for those people, institutions, places, and things that define oneself and give meaning to one's life. Why would one want, desire, long for, lust for a future world beyond one's own lifetime that resembles the one into which one was born? One wants that because that is the world one loves and that is the world one wants to exist for those particular persons that one loves most – one's children, grandchildren, and younger friends, colleagues, and associates. If the world's climate radically changes – as well it might if concerted and effective international action to curb the emission of greenhouse gases does not happen soon – the possibility is very real, perhaps imminent, that the global economy will soon collapse and, following that, the collapse of global civilization. The prospect of a new and irreversible Dark Age looms ominously on the horizon if concerted action to curb the emission of greenhouse gases is postponed much longer. One certainly does not want one's children and grandchildren to live in a world of increasingly violent weather, flooded coastal metropolises, shrunken continents, expanding deserts, desiccated crops – a world ripe for rule by war lords leading predatory gangs, struggling over shrinking stocks of food, energy, and the other necessities of contemporary life that most of us take for granted. And one cringes at the prospect of the destruction of the things that have given one the most joy and reached into the very core of one's being – the art, literature, science, philosophy that we have inherited as the legacy of 5,000 golden years of global civilization. Almost as horrible is the prospect of these currents of global civilization coming to the end of their development, even if somehow they are preserved as relics of a stagnant history. These are all the things that one loves as one's life itself. These are the things one wants to preserve and will gladly work with energy and resolve to perpetuate. This is the ethic of desire, the ethic of love – the post-modern erotic ethic. It stands in sharp contrast to the abstemious, zero-sum, self-sacrificial, guilt-driven – and ultimately ineffectual – ethic of the modern moral monadology.

References

Alaimo, S. 2010. *Bodily natures: Science, environment, and the material self.* Bloomington: Indiana University Press.

Alberts, B., A. Johnson, J. Lewis, M. Raff, K. Roberts, and P. Walter. 2002. *Molecular biology of the cell.* New York: Garland Publishing Inc.

Allen, T.F.H., and T.B. Starr. 1982. *Hierarchy: Perspectives for ecological complexity.* Chicago: University of Chicago Press.

Bennett, J. 2010. *Vibrant matter: A political ecology of things.* Durham: Duke University Press.

Buchman, T.G. 2002. The community of the self. *Nature* 420: 246–251.

Clements, F.E. 1905. *Research methods in ecology.* Lincoln: University Publishing Company.

Coleman, W. 1977. *Biology in the nineteenth century: Problems of form, function, and transformation.* New York: Cambridge University Press.

Collins, R. 1994. *Four sociological traditions.* New York: Oxford University Press.

Grice, E.A., H.H. Kong, S. Conlan, C.B. Deming, J. Davis, A.C. Young, G.G. Bouffard, R.W. Blakesley, P.R. Murray, E.D. Green, M.L. Turner, and J.A. Segre. 2009. Topographical and temporal diversity of the human skin microbiome. *Science* 324(5931): 1190–1192.

Latour, B. (1999). *Politiques de la nature. Comment faire entrer les sciences en démocratie.* Paris: Editions La Decouverte.

Leopold, A. 1949. *A Sand County almanac and sketches here and there.* New York: Oxford University Press.

Lindeman, R.L. 1942. The trophic-dynamic aspect of ecology. *Ecology* 23: 399–418.

Marchesi, J., and F. Shanahan. 2007. The normal intestinal microbiota. *Current Opinion in Infectious Diseases* 20: 508–513.

McIntosh, R.P. 1985. *Background of ecology: Concept and theory.* New York: Cambridge University Press.

Naess, A. 1979. Self-realization in mixed communities of humans, bears, sheep, and wolves. *Inquiry* 22: 231–241.

Odum, E.P. 1969. The strategy of ecosystem development. *Science* 164: 260–270.

Pickett, S.T.A., and R.S. Ostfeld. 1995. The shifting paradigm in ecology. In *A new century for natural resources management*, eds. R.L. Knight and S.F. Bates, 261–278. Washington, DC: Island Press.

Proctor, L.M. 2011. The human microbiome project in 2011 and beyond. *Cell Host & Microbe* 10: 287–291.

Rufli, T., and Y. Mumcuoglu. 1981. The hair follicle mites *Demodex folliculorum* and *Demodex brevis*: Biology and medical importance: A review. *Dermatologica* 162: 1–11.
Sears, C.L. 2005. A dynamic partnership: Celebrating our gut flora. *Anaerobe* 11: 247–251.
Shepard, P. 1969. Introduction: Ecology and man – a viewpoint. In *The subversive science: Essays toward an ecology of man*, eds. P. Shepard and D. McKinley. Boston: Houghton Mifflin.
Tansley, A.G. 1935. The use and abuse of vegetational concepts and terms. *Ecology* 16: 284–307.
Vernadsky, V.I. 1929. *La biosphere*. Paris: Librairie/Feliz Alcan.
Watts, A. 1963. The individual as man/world. *Psychedelic Review* 1: 55–65.
Williams, B. 1981. *Moral luck*. Cambridge: Cambridge University Press.
Williams, G.C. 1966. *Adaptation and natural selection: A critique of some current evolutionary thought*. Princeton: Princeton University Press.
Wilson, C. 2004. *Moral animals: Ideals and constraints in moral theory*. New York: Oxford University Press.
Worster, D. 1977/1985,1994. *Nature's economy: A history of ecological ideas*. San Francisco/New York, Cambridge: Sierra Club Books/Cambridge University Press.
Zimmer, C. 2011. Bacterial ecosystems divide people into 3 types. *New York Times*, April 20.

Chapter 8
Animal Rights and Environmental Ethics

Tom Regan

Abstract The position I favor (the "rights view") prioritizes the moral rights of individuals when it comes to our moral thinking. Some defining features of these rights are explained; reasons for recognizing them in the case of humans are advanced; and arguments for extending them to other-than-human animals are sketched. Several objections are considered, including those that dispute the rights view's alleged inability to explain (1) the amorality of predator-prey relations and (2) our obligations to preserve rare and endangered species.

The position that I favour in ethics ("the rights view") sometimes is criticized because of its alleged inability to address important issues in environmental ethics. As I hope to be able to explain, I believe criticisms of this sort, though understandable, are deficient. When all the dust settles, the rights view grounds important restrictions on our freedom to exploit or destroy the natural world. Granted, some critics want more. In fact, some disparage the very idea of individual rights, viewing it as offering a shallow environmentalism at best, weighed down by antiquated, patriarchal modes of thinking, unequal to the task of plumbing the depths of a deep, bio-centric ecology. I have addressed these issues elsewhere (see, for example, Regan 1991, 1994, and Chapter one, 2001b) and beg leave of doing so again here. Here I begin with a sketch of my understanding of basic moral rights, an understanding first articulated at length in *The Case for Animal Rights* and since amplified and clarified in more recent work (Regan 1994, 2001a, b, 2003a, b, 2004).[1]

T. Regan (✉)
Department of Philosophy & Religious Studies, North Carolina State University, Raleigh, NC 27695, USA
e-mail: regan@ncsu.edu

D. Bergandi (ed.), *The Structural Links between Ecology, Evolution and Ethics: The Virtuous Epistemic Circle*, Boston Studies in the Philosophy of Science 296, DOI 10.1007/978-94-007-5067-8_8,

8.1 Defining Characteristics of Moral Rights

8.1.1 "No Trespassing"

To possess moral rights is to have a kind of protection that we might picture as an invisible "No Trespassing" sign. What does this sign prohibit? Two things. First, others are not morally free to harm us; to say this is to say that others are not free to take our life or injure our body as they please. Second, others are not morally free to interfere with our free choice; to say this is to say that others are not free to limit our free choice as they please. In both cases, the "No Trespassing" sign is meant to protect our most important goods (our life, our body, our liberty) by morally limiting the freedom of others.

Things are different when people exceed their rights by violating ours. When this happens, we act within our rights if we fight back, even if this does some serious harm to the aggressor. However, what we may do in self-defense does not translate into a general permission to hurt those who have not done anything wrong.

8.1.2 Equality

Moral rights breathe equality. They are the same for all who have them, differ though we do in many ways. This explains why no human being can justifiably be denied rights for arbitrary, prejudicial, or morally irrelevant reasons. Race is such a reason. To attempt to determine which humans have rights on the basis of race is like trying to sweeten tea by adding salt. What race we are tells us nothing about what rights we have.

The same is no less true of other differences between us. My wife Nancy and I trace our family lineage to different countries; she to Lithuania, I to Ireland. Some of our friends are Christians, some Jews, some Moslems. Others are agnostics or atheists. In the world at large, a few people are very wealthy, many more, very poor. And so it goes. Humans differ in many ways. There is no denying that.

Still, no one who believes in human rights thinks that these differences mark fundamental moral divisions. If we mean anything by the idea of human rights, we mean that we *have them equally*. And we have them equally regardless of our race, gender, religious belief, comparative wealth, intelligence, or date or place of birth, for example.

8.1.3 Trump

Every serious advocate of human rights believes that our rights have greater moral weight than other important human values. To use an analogy from the card game Bridge, our moral rights are trump. Here is what this analogy means.

A hand is dealt. Hearts are trump. The first three cards played are the queen of spades, the king of spades, and the ace of spades. You (the last player) have no spades. However, you do have the two of hearts. Because hearts are trump, your lowly two of hearts beats the queen of spades, beats the king of spades, even beats the ace of spades. This is how powerful the trump suit is in the game of Bridge.

The analogy between trump in Bridge and individual rights in morality should be reasonably clear. There are many important values to consider when we make a moral decision. For example: How will we be affected personally as a result of deciding one way or another? What about our family, friends, neighbours, people who live some place else? It is not hard to write a long list. When we say, "rights are trump," we mean that respect for the rights of individuals is the most important consideration in "the game of morality," so to speak. In particular, we mean that the benefits others derive from violating someone's rights never justify violating them.

8.1.4 Respect

In a general sense, the rights mentioned above (life, liberty, and bodily integrity) are variations on a main theme, that theme being respect. I show my respect for you by respecting these rights in your life. You show your respect for me by doing the same thing. Respect is the main theme because treating one another with respect *just is* treating one another in ways that respect our other rights. Our most fundamental right, then, the right that unifies all our other rights, is our right to be treated with respect.

8.2 Who Has Moral Rights?

It is one thing to say what moral rights are and quite another to explain why we have them but sticks and stones do not. Given the constraints of space, it will not be possible for me to offer anything like a complete explanation. But permit me to offer a rough sketch of the answer I favour, an answer that relies heavily on what I call a subject-of-a-life.

8.2.1 Subjects-of-a-Life

Earlier we noted some of the many ways that humans differ from one another – in terms of gender, race and ethnicity, for example. Despite our many differences, there are some ways in which all humans who have rights are the same. I do not mean because we all belong to the same species (which is true but not relevant). And I do not mean because we all are persons (which may be relevant but is not true).

What I mean is that we are like one another in relevant ways, ways that relate to the rights we have: our rights to life, to bodily integrity, and to liberty.

For consider. Not only are we all in the world, we all are aware of the world, and aware as well of what happens to us. Moreover, what happens to us – whether to our body, or our freedom, or our life itself – matters to us because it makes a difference to the quality and duration of our life, as experienced by us, whether anybody else cares about this or not. Whatever our differences, these are our fundamental similarities.

We have no commonly used word that names this family of similarities. "Human being" does not do the job (a deceased human being is a human being but is not aware of the world, for example). Neither does "person" (human infants are aware of what happens to them but are not persons). Still, these similarities are important enough to warrant a verbal marker of their own. I use the expression "subject-of-a-life" to refer to them. Given this usage, the author of these words, Tom Regan, is a subject-of-a-life, and so are the people who hear them.

Which humans are subjects-of-a-life? All those humans who have the family of similarities mentioned above. And who might these be? Well, somewhere in the neighborhood of *seven billion* of us, regardless of where we live, how old we are, our race or gender or class, our religious or political beliefs, our level of intelligence, and so on through a very long inventory of our differences.

Why is being the subject-of-a-life an important idea? Because the family of characteristics that define this idea *makes us all equal* in a way that makes sense of our moral equality. Here is what I mean.

As implied in the preceding, human subjects-of-a-life differ in many ways. For example, some are geniuses and others are severely mentally disadvantaged; some are gifted in music while others cannot carry a tune.

These differences are real, and they matter. However, when we think about the world in terms of fundamental moral equality, these differences make no difference. Morally considered, a child *protégé* who can play Chopin *études* with one hand tied behind her back does not have a "higher" rank than a seriously mentally-impaired adult who has never understood what a piano is or who Chopin was. Morally, we do not carve-up the world in this way, placing the Einsteins in the "superior" category, "above" the "inferior" Homer Simpsons of the world. The less gifted do not exist to serve the interests of the more gifted. The former are not mere things when compared to the latter, to be used as means to their ends. From the moral point of view, each of us is equal because each of us is equally a somebody, not a something, the subject-of-a-life, not a life without a subject.

So why is the idea of being the subject-of-a-life important? Because it illuminates our moral sameness, our moral equality.

As subjects-of-a-life, we are all the same because we are all in the world.
As subjects-of-a-life, we are all the same because we are all aware of the world.
As subjects-of-a-life, we are all the same because what happens to us matters to us.
As subjects-of-a-life, what happens to us matters to us because it makes a difference to the quality and duration of our life.
As subjects-of-a-life, there is no superior or inferior, no higher or lower.

As subjects-of-a-life, we are all morally the same.
As subjects-of-a-life, we are all morally equal.

Needless to say, the forgoing does not constitute a strict proof of our rights based on our subjectivity. My intention, rather, has been to explain how our being subjects-of-a-life illuminates (helps us understand) the underpinning of our rights, especially our moral sameness, our moral equality. It should come as no surprise that I think what I have just said about our rights is no less true of the rights of other animals.

8.2.2 Animal Rights

Are any other-than-human animals subjects-of-a-life? Yes, of course. All mammals and birds, most certainly. All fish, most probably. Why? Because (for reasons I have given at length on other occasions and will not rehearse here: see Regan 1983, 2001b, 2003a, b) these beings satisfy the conditions of the kind of subjectivity in question. Like us, they are in the world, aware of the world, aware of what happens to them; and what happens to them (to their body, their freedom, their life) matters to them, whether anyone else cares about this or not. Thus do these beings share the rights we have mentioned, including the right to be treated with respect.

This conclusion (that *these* animals, at least, have basic moral rights) has profound, one might even say revolutionary consequences. Respect for these rights means (among other things) more than cutting back on the amount of meat we eat, or avoiding pale veal, or eating only chicken and fish. It means an end to commercial animal agriculture, whether intensive or free range. We do not respect the rights of cows and pigs, chickens and geese, tuna and trout by ending their life prematurely, however "humane" the methods used. These animals have a right to life no less certainly than we do.

8.3 A Number of Environmentally-based Objections Have Been Raised Against the Rights View[2]

8.3.1 The Rights View and Predator-Prey Relations

Although the main focus of the rights view is duties of justice, there is room within this outlook to include a general duty of beneficence, of doing good for others, not only doing what is just. If (as I believe) we humans have duties of assistance to one another, independent of the demands of justice, there is no reason why duties of the same kind might not arise in circumstances in which animals are involved. For example, suppose a lion is stalking a small child. If we frighten the lion, we may be able to save the child. Since lions are not moral agents, in the sense in which I use this expression, no rights violation is in the offing. But the child almost certainly will be harmed if we do nothing. Should we try to prevent this outcome? Do we have a *prima facie* duty to intervene? It is hard to imagine how a negative answer could be defended. So let us assume (what I take to be true) that we have a *prima facie* duty of assistance in this case.

Next, suppose the same lion is stalking, not a child, but a wildebeest. And suppose, again, that if we frighten the lion, we may be able to save the wildebeest. Since lions are not moral agents, in the sense in which I use this expression, no rights violation is in the offing. But the wildebeest almost certainly will be harmed if we do nothing. Should we try to prevent this outcome? Do we have a *prima facie* duty to intervene? My answer has been, and remains, no. It did not take long for critics (e.g., Ferré 1986) to think that something had gone wrong.

J. Baird Callicott, one of the true pioneers in environmental ethics, is representative. As part of his critique of the rights view, he writes: "If we ought to protect humans' rights not to be preyed upon by ... animal predators, then we ought to protect animals' rights not to be preyed upon by ... animal predators" (Callicott 1989, 45). And not just a little. Callicott insists that the rights view is committed to protecting prey animals a lot. In his words, "Regan's theory of animal rights implies a policy of human predator extermination, since predators, however innocently, violate the rights of their victims" (*Ibid.*).

Whatever else may be true, Callicott clearly overstates his diagnoses when he writes that "predators, however innocently, violate the rights of their victims." Only moral agents are capable of violating rights, and non-human animals are not moral agents. Moreover, and obviously, Callicott moves uncritically from asking what should be done in particular cases, to what should be done as a matter of general policy. And this is important. While we all agree (I assume) that we have a *prima facie* duty to assist the child from the lion, no advocate of children's rights is thereby logically committed to promulgating policies that seek to eradicate every predatory animal under the sun. Why, then, suppose that advocates of animal rights are committed to promulgating such policies because predatory animals harm their prey? Callicott does not say. To tar the rights view with the broad brush of "eradicating wildlife," while it may make for good rhetoric, does not make for good philosophy.

These matters to one side, what does the rights view say about predator-prey relations? To begin with, my position is diametrically opposed to the one Callicott would foist upon me. Instead of advocating a policy of massive intervention in the affairs of wildlife, what we ought in general to do is ... nothing. Here is what I mean and why I think this way.

In my view (see *The Case for Animal Rights*, 1983, 357, 361), our ruling obligation with regard to wild animals is to *let them be*, an obligation grounded in the recognition of their general competence to get on with the business of living, a competence that we find among members of both predator and prey species. After all, if members of prey species, including the young, were unable to survive without our assistance, there would not be prey species. And the same applies to predators. In short, we honour the competence of animals in the wild by permitting them to use their natural abilities, even in the face of their competing needs. As a general rule, they do not need help from us in their struggle for survival, and we do not fail to discharge our duty when we choose not to lend our assistance.

We do not find this same competence in young children. The plain fact is, they cannot take care of themselves and have no realistic hope of surviving, in any circumstances, in the wild or in the home, if we do not help them. To *let children be*, therefore, is not to honour their competence. In general, they do need help from us with their survival skills (whatever these might be). From the perspective of the rights view, therefore, there is nothing in the least bit inconsistent in recognizing duties of assistance to human beings, including human children, that we do not recognize in the case of other animals, including wild animals.

This same point can be made in another way. By my lights (*Ibid.* 103–109), animals are capable of knowing what they want and of acting with the intention of getting it. Because they have these capacities, we can act paternalistically toward them. Roughly speaking (*Ibid.* 107 for greater specificity), paternalistic intervention in their life means taking measures to prevent them from pursuing what they want because, we believe, permitting them to do so will be detrimental to their interests.

When it comes to our obligations to wild animals, the rights view is unapologetically anti-paternalistic. I write: "[T]he goal of wildlife management should be to defend wild animals in the possession of their rights, providing them with the opportunity to live their own life, by their own lights, as best they can, spared that human predation that goes by the name 'sport' [hunting]" (*Ibid.* 357).

In the case of young children, our obligations differ. Someone who placed young children in the woods or on an ice flow, the better to provide them with the "opportunity to live their own life, by their own lights, as best they can," would be judged criminally irresponsible, and rightly so. In general, we act in ways that respect the rights of wild animals by adopting an anti-paternalistic stance, just as, in general, we act in ways that respect the rights of young children if the stance we adopt is paternalistic. From the perspective of the rights view, both stances show equal respect for the rights of both (see for example Everett [2001] defending the rights view against the "predation critique").

8.3.2 The Rights View and Endangered Species

Some environmental philosophers (Callicott is representative) criticize the rights view because of its failure to provide a credible basis for addressing our obligation to preserve endangered species. (For simplicity's sake, I limit my attention to endangered [as distinct from rare] species.) If we set rhetorical excess aside, the logic of the objection is simple. If the rights view fails to provide a credible basis for addressing this obligation, the rights view is not the best theory, all considered.

Although I believe my position is seriously challenged by this line of criticism, and although (for reasons I explain below) I now believe my discussion of endangered species in the past should have been expanded, it is not clear to me that this objection is as telling as its proponents would have us believe. Let me explain.

The rights view restricts rights to individuals. Because species are not individuals, "the rights view does not recognize the rights of species to anything, including survival" (Regan 1983, 359). Moreover, the rights of individuals do not wax or wane depending on how plentiful or rare are the species to which they belong. Beaver do not have lesser rights just because they are more plentiful than bison, and East African black rhinos do not have greater rights than rabbits just because their numbers are declining. How, then, can the rights view address our obligation to preserve endangered species? In the past, I have offered a twofold answer.

First, we have an obligation (*prima facie*, to be sure) to stop human moral agents ("commercial developers, poachers, and other interested third parties" [*Ibid.* 360]) whose actions violate the rights of animals. Second, we have an obligation to "halt the destruction of natural habitat" that makes life for these animals sustainable (*Ibid.* 361). If we succeed in discharging these obligations, my discussion implies, we will succeed in discharging our duty to protect endangered species.

A critic might respond by noting that the rights view fails to do justice to our intuition that we owe something more to endangered than we do to bountiful species. More to East African black rhinos than to rabbits, for example. In view of its insistence on their equal rights, how can the rights view account for this intuition? Here, in rough outline, is the answer I favour.

Compensatory justice is an idea advocates of human justice sometimes employ. A classic example involves past injustice done to members of identifiable groups. For example, although today's descendants of the Miniconjou Sioux who were slaughtered by the 7th US Calvary at Wounded Knee on 29 December 1890 were not alive at the time of the massacre, it is not implausible to argue that they (today's descendants) are owed something because of what happened, not only at Wounded Knee but for many years before and after. Given any reasonable view of history, today's descendants have been disadvantaged because of the massive injustice done to their predecessors. Moreover, what they are owed is something more than what is owed to others of us who have not been disadvantaged in similar ways, for similar reasons. Other things being equal, more should be done for them, by way of compensatory assistance, than what is done for us.

The rights view can apply compensatory principles to animals (the East African black rhino, for example) whose numbers are in severe decline because of past human wrongs (for example, poaching of ancestors and destruction of habitat). Although the remaining rhinos have the same fundamental rights as do members of a more plentiful species (rabbits, say), the duty of assistance owed to the former arguably makes a greater claim on us than this same duty does when owed to the latter. If it is true, as I believe it is, that today's rhinos have been disadvantaged because of human wrongs done to their predecessors, then, other things being equal, more should be done for the rhinos, by way of compensatory assistance, than what should be done for rabbits. In such manner, I believe, the rights view can account for our intuition that we owe members of endangered species of animals something more than what we owe to the members of more plentiful species.

Critics of the rights view can be counted upon to challenge it even after it is augmented by my compensatory argument. In particular, they will point out that the vast majority of endangered species consists of plants and insects, forms of life too rudimentary to qualify as subjects-of-a-life. In their case, because no rights are possessed, nothing can be owed to them for reasons of compensatory justice. Worse (it will be claimed), the continued existence of many of these plants and insects is not necessary to sustain the life of those animals who are subjects-of-a-life. What can the rights view say about our obligation to preserve these endangered species?

What can be said, I think, is what I have said in the past. "The rights view," I write, "does not deny, nor is it antagonistic to recognizing, the importance of aesthetic, scientific, sacramental, and other human interests [in preserving endangered species] ..." (1983, 361) or, more generally, encouraging practices that promote a biotic world at once rich, diverse and sustainable. What the rights view denies, at least given its articulation to date, is that plants and insects are subjects-of-a-life; and it denies as well that these forms of life have been shown to have any rights, including a right to survival. Of course, we may (that is, there is nothing wrong in principle if we do) make great efforts to preserve such life, based on human aesthetic or sacramental interests, for example. But that we may be willing to do this stops well short of establishing that plants and insects have a valid claim against us to do so.

More than a few environmental philosophers in general, including some of the most distinguished among them, will not be satisfied with the environmental implications of the rights view, whether augmented by principles of compensatory justice or not. They will say (in fact some have said – see Rolston 1988) that species have inherent value. And so do ecosystems and the biosphere – which is how we should account for our obligation to save endangered species, including plants and insects, not just "fuzzy mammals." To which (following the lead of the Cuba Gooding character in the movie *Jerry Maquire*) I can only reply: "Show me the argument!" It is not enough to confer inherent (or intrinsic) value on species, ecosystems, the biosphere. One wants a compelling argument for doing so, something that, for reasons I have given elsewhere (Regan 1992), not only has not been done; I believe it cannot be done.

Notes

1. My article is adapted from Chapters 3 and 4 of *Empty Cages: Facing the Challenge of Animal Rights* and from the new Preface to the second edition of *The Case for Animal Rights*.
2. A notable omission is any discussion of Gary Varner's defense of therapeutic hunting (Varner 1998).

References

Callicott, J.B. 1989. Review of Tom Regan, The case for animal rights. In *In defense of the land ethic: Essays in environmental philosophy*, ed. J.B. Callicott, 39–48. Albany: State University of New York Press.

Everett, J. 2001. Environmental ethics, animal welfarism, and the problem of predation: A Bambi lover's respect for nature. *Ethics and the Environment* 6(1): 42–67.

Ferré, F. 1986. Moderation, morals, and meat. *Inquiry* 29(4): 391–406.

Regan, T. 1983. *The case for animal rights*. Berkeley: University of California Press.

Regan, T. 1991. Irreconcilable differences. In *The thee generation: Reflections on the coming revolution*, T. Regan, 125–142. Philadelphia: Temple University Press.

Regan, T. 1992. Does environmental ethics rest on a mistake? *The Monist* 75(2): 161–182.

Regan, T. 1994. The case for animal rights: A decade's passing. In *A quarter century of value inquiry: Presidential addresses of the American Society for Value Inquiry*, ed. R.T. Hull, 439–459. Amsterdam/Atlanta, GA: Éditions Rodopi. Repr. in T. Regan, 2001b, 39–65.

Regan, T. 2001a. The case for animal rights. In *The animal rights debate*, C. Cohen and T. Regan, 127–222, 263–310. Lanham: Rowman & Littlefield Publishers.

Regan, T. 2001b. *Defending animal rights*. Urbana and Chicago: University of Illinois Press.

Regan, T. 2003a. *Animal rights, human wrongs: An introduction to moral philosophy*. Lanham: Rowman & Littlefield Publishers.

Regan, T. 2003b. *Empty cages: Facing the challenge of animal rights*. Lanham: Rowman & Littlefield Publishers.

Regan, T. 2004. *The case for animal rights*. 2nd ed., with a new preface. Berkeley: University of California Press.

Rolston, H. III. 1988. *Environmental ethics: Duties to and values in the natural world*. Philadelphia: Temple University Press.

Varner, G.E. 1998. *In nature's interests? Interests, animal rights, and environmental ethics*. Oxford/New York: Oxford University Press.

Chapter 9
Reconciling Individualist and Deeper Environmentalist Theories? An Exploration

Robin Attfield

Abstract This chapter discusses whether an individualist environmental ethic can be combined and reconciled with an ecocentric or holistic ethic. Versions of individualism include anthropocentrism, sentientism and the variety of biocentrism that I favour. In particular, I consider the value-pluralism advocated by Alan Carter, which seeks, with the aid of multi-dimensional diagrams, to honour a large range of currently held (and supposedly incommensurable) values, including both individualist and ecocentric ones. Carter's description of his own theory accidentally involves contradictions, but even if these are circumvented, there turn out to be problems with endorsing his kind of pluralism, including the absence of reasons or criteria for prioritising values. Arguably, the value of ecosystems depends on that of present and future individuals, and diverse values such as flourishing, achievement, freedom and health can, at least in particular contexts, be prioritised in terms of their value. With the help of arguments adduced by Elinor Mason, I show that, while single-value monistic theories are unsatisfactory, more sophisticated monistic theories for which the values honoured are commensurable are preferable to pluralistic theories such as Carter's for which they are not.

In this paper, I tackle one of the more recurrent issues in environmental ethics, that is, whether an environmental ethic can combine ecocentrism and recognition of holistic values with more traditional approaches such as individualism. While individualism can restrict the purview of morality to human beings (as is the case with anthropocentrism), it is sometimes extended to include sentient creatures; let us call this version of individualism "sentientism." And it is sometimes further extended so as to include non-sentient living individuals as well, a biocentric position. So the question becomes whether an environmental ethic can combine

R. Attfield (✉)
School of English, Communication and Philosophy, Cardiff University, Cardiff, CF10 3EU, UK
e-mail: attfieldr@cf.ac.uk

D. Bergandi (ed.), *The Structural Links between Ecology, Evolution and Ethics: The Virtuous Epistemic Circle*, Boston Studies in the Philosophy of Science 296,
DOI 10.1007/978-94-007-5067-8_9,

ecocentrism with either anthropocentrism or sentientism, or indeed with the individualist biocentrism that I favour. As I shall explain, I have recently been challenged to say why, if at all, this cannot be done.

Although, as Peter Singer has argued, we cannot imagine what it is like for a tree to be harmed, since there are no feelings to imagine, creatures such as trees still have a good of their own and should not be harmed gratuitously, not only for our sakes but also for theirs; or so I (and many others) hold. But sooner than defend this view, I want to consider here an objection to it that would also prove to be an objection to the positions of individualists such as Sober and Regan and Singer as well. Is it not possible and desirable (the objection runs) to combine in some structured manner a plurality of normative theories, and in this way to honour the values stressed by each to a degree that exempts the resulting pluralist theory from the objections to which each is subject? Such a pluralist position will be expounded, discussed and criticized, with a view to shedding further light on value pluralism.

9.1 The Recent Context

I should next explain how I came to be challenged to respond to pluralism in ethics. This challenge arose when Alan Carter reviewed in *Mind* (2001) my 1999 book *The Ethics of the Global Environment*, bringing into the review my more theoretical 1995 work *Value, Obligation and Meta-Ethics*. While a cursory scan could suggest that both books were receiving favourable notices, Carter was in fact arguing that biocentric consequentialism, even if preferable to other monistic theories, fails like all other monistic theories to cope with some of our values, and that a pluralistic theory was to be preferred instead. To do this, Carter purported to find some unpalatable implications of biocentric consequentialism, and claimed that although it could cope with potentially fatal pitfalls such as the Repugnant Conclusion and the Non-Identity problem, its implications actually made it unacceptable to environmentalists, in matters of both population and species preservation (Carter 2001).

Since there were a number of misinterpretations in the review, and since I wanted to correct these and to challenge the supposed unpalatable implications, I published a reply to the review in *Utilitas* (Attfield 2003). In this paper, besides seeking to set the record straight about the matters just mentioned, I argued that ethical pluralism was inherently unlikely to cope with ethical dilemmas because it generates contradictions, and cited with some degree of approval J. Baird Callicott's arguments for a parallel conclusion (Callicott 1990).

To this, Carter has more recently replied in *Utilitas* (March 2005) in "Inegalitarian biocentric consequentialism, the minimax implication and multidimensional value theory." The editor also allowed me a very brief counter-response, which appeared in the same number (Attfield 2005), and which proposes to make that the last word as far as *Utilitas* is concerned, and to close the discussion therewith. In his *Utilitas* paper, Carter makes important distinctions between kinds of moral pluralism. There is (i) theory pluralism, or subscribing to a plurality of normative theories, and also (ii) principle pluralism, or subscribing to a plurality of ethical principles. These are the positions that are challenged by the charge of generating contradictions. But there is also (iii) value pluralism, or subscribing to a plurality of values, and this position is not subject to that charge, involving instead subscribing to a number of distinct, and possibly incommensurable values (Carter 2005, 75). Indeed, each of the major normative theories seeks to maximize one value that it cherishes, and consequently generates one or another "counterintuitive implication by flouting one or more of the other values we hold" (*Ibid.* 75). (Strangely, Carter suggests that biocentric consequentialism gives considerable – and implicitly excessive – prominence to autonomy, and too little to wild species that are inessential to humanity [*Ibid.* 70–71, 76], but I have answered and rejected these charges in *Utilitas* 2005 [see page 86], and need not go over that ground again here.) In order to avoid getting into this kind of position, Carter holds, we "need to give due consideration to each value" (Carter 2005, 75), as his kind of value pluralism supposedly does.

But since "the various values that we hold cannot all be maximally satisfied simultaneously, ... we need to trade them off" (*Ibid.* 76). Carter has come up with an ingenious method of doing this. But before we consider that method, it is worth considering what he supposes that an adequate environmental ethic might look like.

9.2 Carter's Proposed Environmental Ethic

One set of values that an environmental ethic may need to consider, says Carter, includes autonomy, and generally "the way in which we value certain human features, aspirations and projects." These he suggests, "purely for convenience," that we "indiscriminately lump together ... within the category of anthropocentric values" (*Ibid.* 76). But we value other things too. "Many moralists also value, and have been persuaded to value, the interests of all sentient beings. Let us, for convenience, indiscriminately lump all such values within the category of zoocentric values" (*Ibid.* 76). (This passage skates over the possibility that there are animals that lack sentience, but let us ignore this possible problem.) "But – continues Carter – a growing number have also come to value the interests of all living beings. For convenience's sake let us indiscriminately lump together all such values within the category of 'biocentric values'." "Finally", he writes, "some prefer to value the integrity, stability and beauty of the so-called biotic community. A number also value species over and above their members. Let us, for convenience, indiscriminately lump together all such values within the category of 'ecocentric values'" (*Ibid.* 76). To be fair to Carter, the passage quoted is equipped with several footnotes attesting that one or other of these positions is actually held. It should also be noted that nothing more is done to persuade readers that these values really are valuable, let alone that each category is valuable on an irreducible basis. Carter simply proceeds to write as if all this were the case (and thus as if these values are all irreducibly valuable).

To see what kind of pluralist theory Carter regards himself as presenting, it is worth looking at the succeeding passage. First he notes the problem that several ecocentrists have shown that "there appear to be insuperable difficulties in maximally satisfying zoocentric and ecocentric values simultaneously." Here he alludes to Callicott's early paper, "Animal liberation: a triangular affair," and to Mark Sagoff's paper, "Animal liberation and environmental ethics: bad marriage, quick divorce," and seems to accept the relevant common conclusion, but not necessarily the other views argued for in those papers (Callicott 1980; Sagoff 1984). This common conclusion is now contrasted with the easy solution of the problem of relating all these values, which treats anthropocentric values as a subset of zoocentric ones, which are a subset of biocentric ones, which might be held to be a subset of ecocentric ones. As Carter rightly remarks, the various values are not valued on this kind of basis, as if subsets of one another; and as he adroitly adds, ecocentric values are in any case collectivist, while the others are individualist. Thus we cannot integrate respect for all these values on some kind of inclusivist basis. This is why Carter's account of how trade-offs are possible is shortly to be brought on stage (Carter 2005, 76–77).

The reasoning just recounted on Carter's part, however, opens the way to a form of criticism to which Carter's sketch of an adequate environmental ethic may well be vulnerable. For Carter's reasoning shows that he is not treating "anthropocentric," "ecocentric" and the rest as mere meaningless labels, despite his repeated talk of "for the sake of convenience." His conclusions about ecocentric values being different in kind from individualist values are based on the views of actual ecocentrists, etc., and the meanings of these various terms. And this authorizes critical comments to be made about attempts to reconcile values of these heterogeneous sorts. Certainly no suggestion is made by Carter that these terms are being used in any different manner from standard senses, and there is no trace of a definition of any of them, let alone of a new definition. Hence it may be presumed that standard senses are being employed, senses that in fact restrict moral standing and the location of intrinsic value to the relevant classes.

But this means that Carter is trying to reconcile anthropocentric values, which in the standard sense means "values according to which none but human interests matter and all and only humans have moral standing," zoocentric values, which in the standard sense means "values according to which none but animal interests matter and all and only animals have moral standing," biocentric values, which in the standard sense means "values according to which none but the interests of living creatures matter and all and only living creatures have moral standing," and ecocentric values, which in the standard sense means "values according to which either the interests of wholes such as ecosystems and species alone matter or these interests matter independently alongside the interests of one or another set of individuals" (Attfield 2008, 2012). But these four kinds of values are all mutually exclusive and incompatible; not a single pair of these kinds forms an exception or supplies an instance of compatibility. Hence the categories used by Carter make his particular form of value pluralism riddled with contradictions, even if generic value pluralism can be shown not to be subject as such to the charge of generating contradictions in general.

In order to make his method of trade-offs even begin to function, Carter should be using terms that are not by definition contraries: terms such as "autonomy," "freedom from suffering," "health" and possibly "integrity," rather than terms such as "anthropocentric values," "zoocentric values," "biocentric values" and "ecocentric values." Since he actually persists in using the latter terms, we shall need to continue using them, but let us try to do so in a spirit of setting aside the implicit contradictions. For it remains important to see whether a theory of value pluralism of something like the kind that he advocates is possible.

9.3 Carter's Method for Trade-offs

Carter begins by suggesting that the very same environmentalists might refuse to accept some losses to ecocentric value to accommodate anthropocentric value, but also might refuse to accept some losses to anthropocentric value to accommodate ecocentric value (Carter 2005, 77). Let us play along with this suggestion, despite its implausible implication that these environmentalists subscribe to at least two value systems, and seem not to have noticed the incompleteness of either, and the further implausible implication that anthropocentrists would accept anything as compensation for losses to anthropocentric values. What Carter suggests allows him to propose plotting an indifference curve on a two-dimensional graph, with one axis representing anthropocentric values and the other ecocentric values. The resulting indifference curve, he tells us, would be "asymptotic" (*Ibid.* 77). One assumption here is that this trade-off is a little like trading off grapes and potatoes; one would not give up the last of either, but might give up some of one for the sake of some of the other (*Ibid.* 77). It is also assumed that different trade-offs would be accepted for different amounts of each variable, or we would not get a curve at all, as opposed to a single point.

The next issue is how to understand and represent trade-offs between four sets of values. Carter believes this could be done, but for ease of argument and presentation prefers to attempt to present a three-dimensional graph, representing "anthropocentric values along one axis, zoocentric values along another, and ecocentric values along the third" (*Ibid.* 78). [See Fig. 9.1.]

Figure 9.1, Carter tells us, "represents the manner in which we might be indifferent between all points falling on plane 'abc,' and between all points falling on plane 'def.' But we would prefer all points falling on 'abc' to any falling on 'def,' given that all points on the latter plane are closer to the origin than any points on the former" (*Ibid.* 78). So the assumption here is that we have multiple values (values of at least three kinds), and prefer some satisfaction of all of them to higher levels of satisfaction of some combined with negligible or no satisfaction of one or two of the others. If we were relating "values" such as freedom, health and security, this might be a cogent approach.

What Carter may be inviting us to consider is whether, if we set aside categories such as anthropocentric, zoocentric and ecocentric, his multi-dimensional

Fig. 9.1 From Carter, *Utilitas*, 17, 1, March 2005

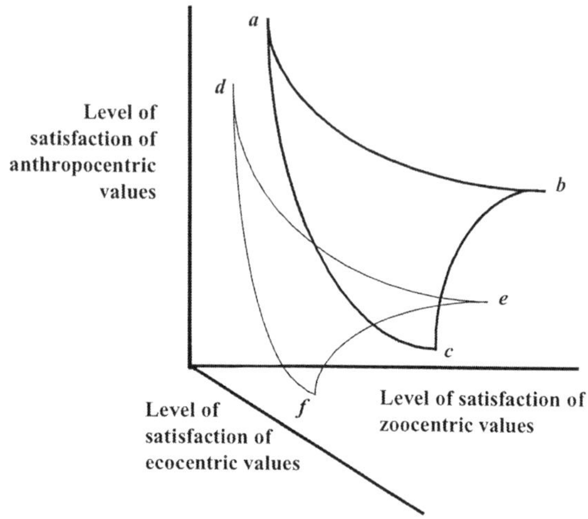

indifference curves are a reasonable way of relating values such as human autonomy, animal welfare and ecological integrity. This approach might well be resisted not only by anthropocentrists, who might well refuse to accept any gains for non-humans as compensation for losses to humanity, but also, for example, by rights theorists, who might hold that rights theory already strikes a proper balance between human rights and animal rights, without needing to be weighed against external factors. This approach would certainly be resisted by biocentric consequentialists committed to a theory of degrees of intrinsic value, for on this view all the different valuable items that need to be considered can be compared in terms of their value and/or disvalue. Probably other kinds of consequentialists would maintain this too.

However, Carter has a further card to play. For not "all points on either plane 'abc' or plane 'def' may represent possible outcomes. The frontier of all possible outcomes could be constituted by plane 'αβγ' as in Fig. 9.2." [See Fig. 9.2.] (Carter actually declares that it *is* so constituted, but since he cannot know this, I am interpreting him as conveying that this is a possibility) (*Ibid*. 79).

Whereas the planes 'abc' and 'def' are concave, plane 'αβγ' is, as we learn from Fig. 9.3, convex, and meets plane 'abc' at just one point, point 'T,' or so Carter claims (*Ibid*. 80).

Thus 'T' represents the only possible outcome that is also desirable on balance once our values have been traded off, and is therefore the outcome that we ought to aim for. Moral pluralism, it emerges, "can generate determinate moral answers" (*Ibid*. 79). If we allow Carter his method, and ignore the framework of contraries that he uses to set it up, then his charts do seem to demonstrate this, even though agents could seldom be in a position to know what any of these planes were like, or to know whether the possible intersects at all with the overall desirable. Indeed, there must be considerable doubt about whether it ever would so intersect, and

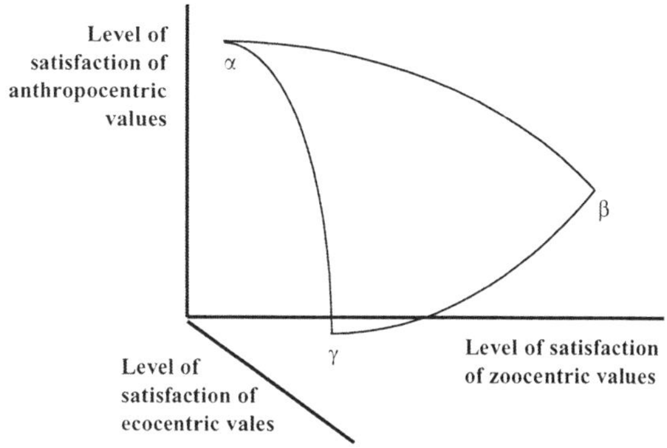

Fig. 9.2 From Carter, *Utilitas*, 17, 1, March 2005

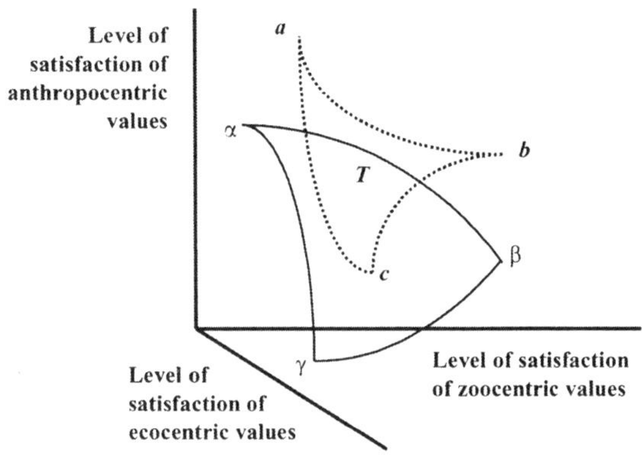

Fig. 9.3 From Carter, *Utilitas*, 17, 1, March 2005

whether this could be known, and again whether if it did intersect there would not usually be a whole arc of options to choose between, rather than a determinate point. Granted that agents would simultaneously be struggling with a sea of probabilities, uncertainties, risks and dangers, the chances that they would ever be assisted by such multi-dimensional analysis seem slender. But I do not intend to pursue these points further here, in order to reflect instead on value pluralism and rival theories. I turn to Carter's comments on monistic theories in the next section.

Here it is worth inserting that Carter envisages adding several further dimensions to his theory. To use his own words, "A fully adequate environmental ethic would need to incorporate what is of value in each theory – namely the values each prioritizes – and successfully combine them" (*Ibid.* 81). In particular, he wants to introduce and trade off against other values not only total human welfare but also average welfare (contrary to the views of those of us who regard average welfare as a derivative value, if a value at all) and, again, its distribution, and, there again, "rights violations and the number of beings with interests who stand to benefit greatly from them." (Indeed, Carter adds that it is this aspect that makes his own views not to be purely consequentialist) (*Ibid.* 81). Here Carter cross-refers to another paper of his in which such trade-offs are set out in greater detail (Carter 2002). The need to take all these values into account explains, he claims, why "all monistic theories are bound to be inadequate," and "why a truly acceptable environmental ethic, as with any acceptable moral theory, will need to be a pluralist one" (Carter 2005, 81). And with these words, Carter seems to be claiming superiority for value pluralism not only over biocentric consequentialism, but equally over sentientist or zoocentric consequentialism, over anthropocentric consequentialism, and at the same time over deontological theories concerned to minimize rights violations, and over Kantian theories too. Indeed, at one point he claims its superiority over Rawlsian theories too, recognizing that they are not monistic, but holding that, granted their lexical ordering of values, they are defective in being unable to enjoin outcome 'T' (*Ibid.* 80).

However, even if these claims were unproblematic and proved to be vindicated at the theoretical level, Carter's theory would make most if not all decision-making extremely complex and contentious. Thus it could be best to stay with more conventional approaches, such as ones that seek to maximize well-being or quality of life among both humans and non-humans, until they are actually shown not to do the work required of them. Yet Carter's stance involves a challenge to all monistic theories, at least at the level of theory, and so it is worth further investigating whether, at least in theory, value pluralism is superior.

9.4 Pluralism and Monism

Just after introducing Fig. 9.3, Carter tells us how he believes monistic theories fare in terms of their outcomes when subjected to multi-dimensional analysis. Here are Carter's words:

> But the outcome enjoined by any monistic theory, in maximizing one value regardless of the rest of the values we hold, will be represented by a point on 'αβγ' that is close to one of the axes. Hence, the outcomes enjoined by monistic theories are bound to strike anyone with a richer sense of values as morally unacceptable, given the practicability of obtaining an outcome that better satisfies her particular combination of values, such as that represented by point 'T' (*Ibid.* 80).

Monism, then, is here supposed to be seeking to maximize one value among the others that we hold, for example, autonomy or animal welfare or justice or equality. This is why it is thought to select a point close to at least one of the axes, for monism supposedly insists on maximizing one value, and through refusing to accept compensation in exchange for sacrifices to such a maximizing project, selects outcomes that will often happen to satisfy other values to a very slight extent indeed. Of course, a monist may attain a very high level of satisfaction of her favoured value, if that is possible, and in making no mention of this, Carter could be held to dismiss monism too readily. But his remark about the views of people with a richer sense of values could well seemingly stand up; and if we concede that the plane 'αβγ' faithfully represents the frontier of possible outcomes, then we automatically forfeit the right to claim that there are any practicable outcomes beyond it. However, what this discussion brings out is that we should not accept that 'αβγ' faithfully represents possible outcomes in all possible worlds in the first place, as opposed to representing a range that would in some possible world be the range of the possible.

So are Carter's interpretations of monism fair to those he regards as monists? They could be fairly applied, it seems to me, to hedonists who seek simply to maximize pleasure and the absence of pain. For it is implausible that values such as autonomy and achievement are adequately recognized within hedonism, however ingenious hedonists may be in reducing these other values to pain or its absence or to pleasure. But it is much less obvious that Carter's remarks apply to theorists such as Derek Parfit, who write about maximizing whatever-makes-life-worthwhile, or to George Edward Moore, who urges us to maximize intrinsic goodness, but holds that a plurality of things are intrinsically good. Similarly, theories such as my own, which commend maximizing the balance of intrinsic value over disvalue, but locate intrinsic value in different degrees in different sources of value, could also elude Carter's adverse comparisons. Indeed, by now it is difficult to tell whether they are, in Carter's terms, monistic or not.

All of this raises large issues about how Carter defines "monism." Thus if monism is restricted to theories seeking to maximize just one value from among the range of values that most people recognize (let us call this "exclusive monism"), then few of the traditional theories of normative ethics are monistic. (Nor, it might be added, do we need multi-dimensional indifference curves to expose such exclusive monist theories as inadequate.) In the passage just cited, Carter does seem to be using "monism" in this exclusive sense; and this might allow theorists such as myself to reject the ascription of monism in his sense, or even to claim to be pluralists, if of a different sort from Carter.

But in holding that most theories of normative ethics are monistic, and that these include anthropocentric, zoocentric, biocentric and ecocentric theories, Carter seems also to employ a broader sense of "monistic." He seems to include among monistic theories (but this is conjectural) ones that recognize more than one value (autonomy and health, for example), but claim that rational preferences are possible between them, or rather between conflicting options in cases where these values are in potential conflict, and where neither can be satisfied without some sacrifice of the other. But he probably also holds that monistic theories do not recognize some of the range of values that one or another set of theorists purport to recognize. Thus anthropocentrists neglect intrinsic value in the lives of non-human creatures, sentientists deny intrinsic value in the lives of non-sentient creatures, and biocentrists, such as myself, deny *intrinsic* value of the kind that ecocentrists purport to recognize in ecosystems and in species. And if this is what Carter means by monism (let us call this sense "inclusive monism"), then I (and probably many others) will be correctly depicted as monists (and inclusive monists at that), despite recognizing a plurality of locations of value (in autonomy and health, for example, for the same examples will serve again).

Yet it should at once be remarked that affiliation to inclusive monism need not commit a theorist to selecting an outcome on Fig. 9.3 closer to any of the axes than point 'T.' For an inclusive monist may hold that the satisfaction of human interests such as autonomy has a degree of intrinsic value smaller or greater than or equal to that of the health or wellbeing of an animal or of a tree, and may reach outcomes not by seeking to maximize just one of these sources or locations of value, but by weighing possible outcomes in terms of their degree of value, or the balance of value over disvalue involved. Such a theorist can be seen as engaged in comparisons and appraisals in which not all the kinds of value are maximized, but in which more than one are honoured. Hence such a theorist need not prioritize one kind of value at the expense of all others, despite Carter's claims that this is the invariable tendency of monism. This being so, the inclusive monist can escape Carter's claim that her chosen outcome in the terms of Fig. 9.3 is bound to be less satisfactory than that of the value-pluralist, and relatedly his claim that it shows value pluralism to be superior to the various kinds of monism (*Ibid.* 81).

Another significant difference between monism and Carter's kind of pluralism should now be noted. For Carter's pluralism seems committed from the start to regarding what he calls "the various values that we hold" as one and all of them values to be separately honoured, as if each of them (as he puts it) "continually exercises its pull" independently of the others, and as if none of them might be derivative values, or not values at all. For example, the value of ecosystems is assumed to be an independent value to be taken into account, without consideration of the view of individualists that the value of such systems, important as it is and remains, is dependent on the value of the individuals (present and future) whose existence these systems make possible. By contrast, monisms of every stripe draw the line somewhere, and reject some of the claims made about the range of independent values. (Those ecocentrists who recognize value solely in collectivities and not at all in individuals are here just as monistic as their individualist opponents.)

Thus Carter's value pluralism is in this regard broader and more tolerant than any of the stances of the theorists just mentioned; according to Carter, the whole proposed spectrum of values are to be honoured as both genuine, independent and deserving of recognition. But this aspect of his theory is as much a danger as an asset. For if he is wrong about the independent value of ecosystems, but proceeds to factor this into his multi-dimensional decision-making procedure, then every outcome, verdict and judgement emerging from that procedure will be skewed. And this is a possibility with regard to every value that he endorses; thus if biocentrists are wrong about there being intrinsic value in the flourishing of trees, then Carter's value pluralism is equally in error, and so on. (It could further be asked how Carter could resist including within his range of values the values that, say, moon-worshippers might advocate; if seleno-centrism became widespread, would he not be obliged to include seleno-centric values within his multi-dimensional analysis? If not, on what basis could he justify excluding them?)

What this suggests is that it is possible to construct an ethic that combines ecocentric values with individualist values (including individualist values of the anthropocentric, sentientist, and biocentric kinds), but that the resulting ethic will remain implausible unless it can be equipped with some kind of rational defence. The various monistic theories each put forward some kind of defence for drawing the boundary of moral considerability and of the location of intrinsic value where they do, and could each be held to be in this respect preferable to Carter's value pluralism. Carter, for his part, assumes that the grounds that the various monisms supply for drawing the line where they do are one and all misguided. But this places the onus on him to show that this is the case, and why it is. Otherwise there are plentiful grounds for holding that Carter's whole-spectrum approach is vulnerable, and that the multi-dimensional decision-procedure based on it is not only unduly complicated but probably profoundly misleading.

At this point, it is salutary to remember the exclusionary meanings attaching to anthropocentrism, sentientism, biocentrism and, for that matter, ecocentrism of the purely holistic kind. While each of these theories can be held in inclusive versions (to use the terminology introduced earlier) through recognizing a range of values, each of these has some kind of coherence because it monistically affirms that moral standing and intrinsic value are located only where it says, and not where rival theorists suggest. We have already seen the problems of commitment to potential contradictions that Carter generates for himself by seeking to combine the values of all these contrary kinds of theory. The question now is whether, even if these labels are set aside, combining all these monisms in a whole-spectrum pluralism generates a theory that is defensible, and that anyone would be motivated to hold.

9.5 Elinor Mason on Monism, Pluralism and the Comparison Thesis

In this final section, I want to relate the above issues to a paper in which Elinor Mason defends monism against foundational pluralism. Mason's paper, "The High Price of Pluralism", is unpublished, and so it is inappropriate to mention more than its main thrust.

By foundational pluralism, Mason means theories that represent values as irretrievably plural and beyond comparison. But if morality is not to be an impossible enterprise, comparisons must be possible. Hence foundational pluralism is a wildly implausible position, and monism, the kind of value-theory that allows of comparisons, is to be preferred.

Mason illustrates her thesis with plentiful examples drawn from the history of ethics, but it is more important to make it clear that she is not defending what I have called exclusive monism. She recognizes that several kinds of thing may be valuable, just as inclusive monists (in my terminology above) are prepared to do. For example, knowledge and friendship were both examples of goodness for Moore. This being the case, Mason suggests that we call knowledge and friendship "sources of value" (as I have occasionally done above) or "non-basic values" rather than simply "values"; for the monist invariably wants to go on to hold that there is something by virtue of which they can be compared, and for that we may reserve the term "value" (although Moore employed the term "goodness"). And here the monist is, according to Mason, fundamentally right, in virtue of what she calls "The Comparison Thesis." This is the thought that "if A is better than B, it must be with respect to something. If there is no relevant feature in terms of which to compare A and B, then A and B ... cannot be compared at all" (pp. 3 and 6 of unpublished manuscript). Thus someone who recognizes several values (which we should rename "sources of value") must hold that there is something in virtue of which comparisons are possible, or, in other words, value.

In expounding her position, Mason explains that comparisons are not achieved simply by expressing either preferences or approvals. Some ground or basis must be available on demand to justify any rational comparison. (Relevant criteria will sometimes be plural, and may need to be combined or blended, but the monist can readily cope with such complexity, holding that various combinations or proportions of desirable features can be ranked and are better or more valuable than others.) Here, and in her general defence of value-monism (which is much more detailed and sophisticated than I have space to mention here), she seems to me correct (although when she comes to list possible sources of value, her inclusion of "the environment" [pp. 32–33 of unpublished manuscript] seems to lack specificity). What is less clear is how her arguments should best be applied to Carter's value pluralism.

Carter could be held not to be a foundational pluralist in Mason's sense, because he allows the various values that he recognizes to be compared in multidimensional indifference curves and valuations. On the other hand, none of the verdicts that generate the planes that form these indifference curves seem to involve any basis of comparison; rather, it is held that "we" will select the plane 'abc,' prefer it to the plane 'def,' etc. Admittedly, there is some rational basis for these preferences; more rather than less of what we value is preferable, and those outcomes are preferable which respect all the values that we hold rather than just one or some. But is this, in the end, a form of rational comparison, as opposed to a systemizing of preferences (either of one person or, if we are lucky, of several people with the same preferences)? Remember that the axes of the diagrams represent autonomy or flourishing or ecosystem integrity or the like; no attempt is made to chart rational preferability or (as we might re-express that concept) value.

But reflection on rational preferability is surely just what is needed, and it has been seen to be lacking from the kind of whole-spectrum pluralism that Carter advocates. Within Carter's system of thought, it is his very inclusiveness that takes the place of such reflection. Thus Mason's stress on the need for rational comparisons to have a clear basis may be just the kind of corrective to Carter that is needed. This granted, inclusive monists are free to recognize a wide range of environmental and other values, or rather sources of value, and to reason about which of them embody value intrinsically as opposed to derivatively, and to recognize a variety of degrees of value, and to attempt to arrive at ethical judgements and principles accordingly. What facilitates all this is the reasoned approach common to monisms but effectively rejected by pluralisms that deny comparability. It is because value pluralism of Carter's kind by-passes all this that its procedures and deliverances fail to show how rationally to combine multiple normative theories, and thus how rationally to combine ecocentrism with individualism.

References

Attfield, R. 1995. *Value, obligation, and meta-ethics*. Amsterdam/Atlanta, GA: Éditions Rodopi.
Attfield, R. 1999. *The ethics of the global environment*. Edinburgh/West Lafayette: Edinburgh University Press/Purdue University Press.
Attfield, R. 2003. Biocentric consequentialism, pluralism, and the 'minimax implication': A reply to Carter. *Utilitas* 15(1): 76–91.
Attfield, R. 2005. Biocentric consequentialism and value pluralism: A response to Alan Carter. *Utilitas* 17(1): 85–92.
Attfield, R. (ed.). 2008. *The ethics of the environment*. Farnham/Burlington: Ashgate.
Attfield, R. 2012. *Ethics: An overview*. London/New York: Continuum.
Callicott, J.B. 1980/1989. Animal liberation: A triangular affair. *Environmental Ethics* 2: 311–338. Repr., *In defense of the land ethic: Essays in environmental philosophy*, J.B. Callicott, 15–38. Albany: State University of New York Press.
Callicott, J.B. 1990. The case against moral pluralism. *Environmental Ethics* 12: 99–124.
Carter, A. 2001. Review of Robin Attfield, The Ethics of the global environment. *Mind* 110: 149–153.
Carter, A. 2002. Value-pluralist egalitarianism. *Journal of Philosophy* 99(11): 577–599.
Carter, A. 2005. Inegalitarian biocentric consequentialism, the minimax implication and multidimensional value theory: A brief proposal for a new direction in environmental ethics. *Utilitas* 17(1): 62–84.
Mason, E. The high price of pluralism. (unpublished paper, prepared 2004).
Sagoff, M. 1984/1993. Animal liberation and environmental ethics: Bad marriage, quick divorce. *Osgoode Hall Law Journal* 22(2): 297–307. Repr., *Environmental philosophy: From animal rights to radical ecology*, eds. M.E. Zimmerman, J.B. Callicott, G. Sessions, K.J. Warren, and J. Clark, 84–94. Englewood Cliffs: Prentice Hall.

Chapter 10
Two Philosophies of the Environmental Crisis

Catherine Larrère

Abstract One of the most important – and most disturbing – characteristics of philosophical reflection on environmental questions is that there are, in reality, two separate issues involved. One refers to a philosophy of nature and the other to a philosophy of technology. This has led to two forms of well-established and clearly argued reflection, each with its own debates. These two currents have developed independently of each other, and continue to do so, as if the other did not exist. But this duality is no longer tenable. Due to the generalization of the environmental crisis and the emergence of new technologies, it has become impossible to treat nature and technology separately. This paper is thus an attempt at a synthesis of these two fields of environmental ethics.

Soil erosion, pollution of all kinds, species extinctions, the greenhouse effect and global warming, holes in the ozone layer – all of these well-known phenomena form part of what has been called the environmental crisis. This crisis has drawn great attention from different fields of knowledge, philosophy among them. Much has been written about environmental philosophy – and more especially environmental ethics – so much so that some people grew afraid that these new philosophical developments could threaten well-established certainties. At the beginning of the 1990s, Luc Ferry wrote a book, *The New Ecological Order* (1995 [1992]), in which he contended that giving rights to nature was tantamount to destroying human rights, and that green movements were prone to becoming fascists.

Ferry's attacks, it seems to me, backfired, mostly because he chose the wrong target. In targeting what he called "deep ecology" (making a scarecrow out of it), while he had in mind European environmentalism (and more parochially still, the

C. Larrère (✉)
Department of Philosophy, University of Paris 1 (Panthéon-Sorbonne),
75005, Paris, France
e-mail: c.larrere@wanadoo.fr

D. Bergandi (ed.), *The Structural Links between Ecology, Evolution and Ethics: The Virtuous Epistemic Circle*, Boston Studies in the Philosophy of Science 296,
DOI 10.1007/978-94-007-5067-8_10,

French Green Party), he seemed to be unaware that, especially in Europe, a large part of philosophical thinking about the environment is not so much concerned with nature as with technology. One of the most striking, and perhaps the most disturbing, features of environmental philosophical thought is that there is not one, but two current philosophies of the environment. One relies on a philosophy of nature, the other one relies on a philosophy of technology. Each ignores the other. They have been developed out of different contexts, and deal with different problems. But they can no longer remain separate. Some attempt has to be made to link them together.

Hence we intend to: (1) briefly present both approaches; (2) explain why they cannot remain separate any longer; and (3) ask whether and how it is possible to build a bridge between them: we will do this by way of a critical examination of Bruno Latour's attempt to elaborate a general response to the environmental crisis in *Politics of Nature* (2004 [1999]).

10.1 Philosophy of Nature or Philosophy of Technology

In a relatively recent issue of the *Journal of Agricultural and Environmental Ethics* (2003), James B. Gerrie documents this divide in approaches to environmental problems, clearly distinguishing between what he calls the "White hypothesis" and the "Ellul hypothesis." The difference between the two hypotheses stems from the respective views of what the environmental crisis is related to.

1. Lynn White Jr is famous for having published a paper in 1967 on "The Historical Roots of Our Ecological Crisis," where he charges Christianity (or Judeo-Christianity) with responsibility for the environmental crisis because it enhances humanity's contempt for nature. In Judeo-Christian religion, nature is but an instrument that God puts in the hands of men. White's paper can be seen as the starting point of an interrogation about "what is wrong in man's relationship to nature," an interrogation that gave birth to environmental ethics as an ethical concern about nature, as a way of taking intrinsic value in nature seriously.

   This issue of the intrinsic value in nature is mostly what environmental ethics is about in the English-speaking world, and this has led to distinguishing between approaches with different focuses: biocentric, ecocentric, anthropocentric. All of them are related to a philosophy (Gerrie would say a metaphysics) of nature, and develop an ethics of respect for nature. The general idea is that if we change our relationship to nature by taking into consideration its moral dimension, we will behave both rightly and usefully.
2. The "Ellul hypothesis" is completely different. The environmental crisis is related not to our (wrong) relationship to nature, but to our capacity to assess and to master our technological activities. Jacques Ellul seems indeed to have been the first to elaborate the thesis of the "technological system" (Ellul 1964 [1954]), what is called in French "the autonomy of technology" and in English

"technological determinism," the general idea being that we are now unable to master our technology. It has become an autonomous, self-developing process, which confronts us as a necessity. Technologies form a system, and this system holds us prisoner. The challenge is to master this system, to develop the ability to limit or steer the technological process.

In this approach, the question is not so much our respect for nature as our responsibility towards a technological process that we initiated but no longer fully master.

Because there are two ways of explaining the environmental crisis, there are two ways of addressing it, and two different ethics: an ethics of respect for nature (which we will call the "naturalistic environmental ethics") and an ethics of responsibility (which we will call the "ethics of technology").

Each set of ethics ignores the other. James Gerrie performed a survey of textbooks and encyclopedias of naturalistic environmental ethics, and concluded that the "Ellul hypothesis" was almost never mentioned. I presume that a similar survey of technological environmental ethics would yield the same result.

In one way, this reciprocal ignorance is no surprise. The approaches arise out of very different contexts, as responses to very different problems.

1. The first approach is to be found in the English-speaking countries, mainly in the former British colonies (North America, Australia, New Zealand). It is strongly related to the way that the pioneers transformed and destroyed the natural environment that they had found. Whereas in Europe taming wild nature had been a very long process, a relation between people and nature going back and forth, for centuries and centuries, in America it took less than a century for the pioneers to destroy the wilderness, and to become ashamed of having done so. Naturalistic environmental ethics originated in the desire, during the nineteenth century, to preserve this threatened wilderness.
2. European environmental concern mostly originated later on, out of worry about twentieth-century technology, and mostly nuclear technology. This could be one reason why environmental concerns in Europe most often find a political rather than ethical expression: it has to do more with political collective decision-making than with individual behaviours.

Having emerged in different contexts, the two different ethics (or concerns) are facing different questions: the preservation of nature, on the one hand, and risk-avoiding policies, on the other. These two sets of ethics are not even really dealing with the same nature. Nature preservation is about a very visible and sensible nature, a nature we see, feel, and love. As far as scientific knowledge is required, it is ecology. Risk assessment is about a much more abstract nature, about physical, chemical, biological processes, a mostly invisible nature, a nature that we can access only through very sophisticated instruments.

So naturalistic environmental ethics and technological assessment are two separate areas that can coexist without competing or overlapping – or rather which could once coexist. For this is no longer the case.

10.2 A Non-sustainable Duality

To show this, I will take an example. Fire is part of natural processes, and can be seen as an example of the disturbances that play an important part in natural successions. Many plant species depend on fire to survive their competition with other plants. So this is the reason why most nature preservationists (and people in charge of preserved natural areas) think that the right thing to do (for both ecological and ethical reasons) is simply to let forests burn in wilderness areas. This is what happens every summer in North America.

On the other hand, burning forests release a huge quantity of greenhouse gases. When such fires occur in very large areas, these effects cannot be overlooked or ignored. This is so much so that, at the Kyoto conference, in assessing global warming, and how each country could contribute to facing the crisis, the fact that some countries had very large wooded areas (carbon wells) was taken into account in balancing the release of gases. So letting forests burn not only adds to the greenhouse effect, but also diminishes the volume of captured carbon.

So this is a case of conflicting ethical indictments: let the forest burn, out of respect for nature, or stop the fire to master the greenhouse effect. This can be seen as a consequence of the globalization of the environmental crisis: it calls into question nature preservation solutions that, though well-founded, are based on local considerations.

There is yet another reason for calling into question the separation between the two ethics: it is more and more difficult to tell apart the natural and the artificial, as the border between them has definitely blurred.

(a) There is no longer any true wilderness, that is, a nature which is completely apart from man, completely free from human transformation. Even in (so-called) wilderness areas, where (according to the Wilderness Act) man is only a temporary visitor, there are too many visitors, and they have become the main threat to preserving the wilderness.
(b) How can one tell what is artificial and what is natural? Take GMOs, for instance. They are highly artificial: they are the result of human design (they are intentional, that is), and they cannot exist without very sophisticated technological instruments and scientific knowledge. At the same time, they are very natural: they live and reproduce by themselves, without human intervention (and these are criteria for naturalness), which is the very reason why many people fear their environmental and agronomical effects: once they have been released into nature, one cannot call them back. They live their own lives.

In such a situation, one can see that the two approaches are no longer coexisting, they are competing, and the technological approach is winning over the naturalistic one. This is especially clear with the worldwide success of the ethics of sustainable development. Not only is sustainable development obviously anthropocentric (to speak the language of naturalistic environmental ethics), since it is concerned with future generations and views nature solely as a resource to safeguard for future generations, but even more, sustainable development means the victory of the conservation of resources over the preservation of nature, of Pinchot's heritage over Muir's heritage.

A close examination of the proceedings of the United Nations Earth Summit in Rio de Janeiro (1992) should show how preservation objectives were superseded by conservation objectives (especially by studying the Convention on Biological Diversity and its preparatory documents). I will simply refer to something symbolic: it was while adopting the sustainable development policy that the former International Union for the Preservation of Nature (IUPN) was changed into the International Union for Conservation of Nature (IUCN). In the controversy between Muir and Pinchot, between preservation and conservation, Muir certainly won as far as the wilderness movement and naturalistic environmental ethics were concerned. But with respect to a more global environmental policy, Pinchot is winning out over Muir. It is sufficient to recall Pinchot's saying: "The first great fact about conservation is that it stands for development."

I would just like to add the hypothesis that the victory of Pinchot, or of conservation, was made all the easier, because the idea of conservation was fueled by the critical philosophy of technology. And reference to technological environmental concern more or less erases any reference to nature. So it is not only that conservation is winning out over preservation, but also that technological concern is winning out over concern for nature.

Is this victory a cause for rejoicing? The technological approach has some arguments in its favor. It can be argued that the technological approach is more pragmatic, less metaphysical than the naturalistic one (this is Gerrie's argument). It can also be argued that mastering our technology is a prior condition for transforming our relationship with nature. Why advocate respect for nature if we are not able to limit our technological power?

But is this victory a true victory? Can we really make do without any reference to nature? To answer this question, it is interesting to study Latour's proposal, in *Politics of Nature*, of an environmental policy. For, unlike most technologically concerned environmentalists, he does not ignore the naturalistic approach completely. He directly attacks it. He tries to argue that to promote environmental concern, one must get rid of nature.

10.3 Nature or Non-humans?

Politics of Nature is an ill-named book, for Bruno Latour's proposal to environmentalists and green parties is that, if they want to successfully deal with environmental problems, they must get rid of nature. The first chapter is entitled: "Pourquoi l'écologie politique ne peut pas conserver la nature" (a deliberately ambiguous sentence, which can mean that "political ecology" is not about nature preservation, nor even conservation, but also, more generally, that nature is something we must forsake). Latour is advising environmentalists that they should base their policy not on naturalistic concerns (coming from conservationists or ecologists/scientists, that is), but rather on a social and political treatment of scientific controversies.

Latour's basic idea is that nature, far from being part of the solution, as has generally been thought so far, is actually part of the problem. The generally received idea is that, as soon as governments have become aware of the seriousness of the environmental crisis, they have (as in France) created new ministries, or agencies, of the Environment, or of Ecology. This has been seen as extending the political sphere to include natural problems which, hitherto, had been the concern of scientists; it has been seen as a way of bringing nature into politics. Not at all, objects Latour. To do this leads only to perpetuating the problems, which are linked with the function of nature in political and social discourse. Considering the way that nature is a part of our mental frameworks, speaking of nature has two important consequences:

1. Nature is what pertains to scientific inquiry, and what gives the scientist the authority to speak. Speaking of nature, and from nature, the scientist speaks the truth, the one and only truth. This is the way that political problems implying scientific expertise used to be dealt with; the expert speaks, and the politician listens. Nature is what scientists use to reduce politicians to silence.

   But this is not what the environmental crisis teaches us. The environmental crisis is not about scientific certainty. It is about scientific controversy. Be it the bovine spongiform encephalopathy (BSE) crisis, global warming, the erosion of biodiversity or other, these are all issues involving scientific controversy. The environmental crisis has made scientific controversies matters of public knowledge and debate. Hence Bruno Latour (whose main study is of scientific controversies in their social context) advises environmentalists that "écologie politique" should be about joining scientific controversy with political debate. This explains the subtitle of his book: "How to bring the sciences into democracy."
2. Nature, as part of our mental framework, is a principle of classification, or categorization. Nature is to be found in general oppositions such as nature/culture, or nature/artifact, or nature/society. These dichotomies are used to classify objects, or entities, that belong either to nature or to culture. This way of dividing objects between large categories, nature being always on one side, is what Latour, in a previous book, *We Have Never Been Modern* (1993 [1991]), calls the "modern constitution": a divide that is not given anywhere, but which was constructed, in the modern era (beginning with the seventeenth century) mostly around modern science (Galileo and Descartes). It has never worked completely, but it is working less and less. The environmental crisis is calling the "modern constitution" into question. It comes with the proliferation of "hybrid objects," that is, objects that are both artificial and natural, and these are the main objects of the environmental crisis: global warming is both man-made and natural, as are GMOs, etc. Or, as long as the modern constitution rules our world (or our way of viewing the world), such hybrid objects are merely

invisible, because they cannot be classified under one of the two opposite poles of the constitution: they are neither natural, nor artificial. Hence Latour speaks of the environmental crisis as the "revolt of the objects." They do not want to be treated any longer as "objects" (as opposed to "subjects," another dichotomy of the "modern constitution"), as clean-cut, easy-to-grasp things. Latour speaks of them as being "hairy."

The solution proposed by Latour is to repudiate the modern constitution, which assigned objects to nature, and according to which communities were made only of subjects, of humans, and to create a new community that includes both humans as well as "non-humans." The model of such a community is to be found in the sciences, for the sciences give rise to social communities, which unlike the usual communities (and especially political communities) are able to include objects, or so-called "objects," now called "non-humans." This refers to Michel Serres' *The Natural Contract* (1995 [1990]), in which he shows how science is able to make objects bear testimony about themselves, in much the same way that people can bear testimony in a judiciary trial.

So Latour argues that political communities should elaborate procedures asking scientists as well as political representatives to examine, select, and assess the new non-human members of the communities.

When one looks at the lists Latour gives as examples of such "non-humans," one finds artificial as well as natural members: Latour lists a lion, asbestos, global warming, a river, a herd of elephants, GMOs, a prion, etc. Such a list verifies that the nature/artifact divide has truly been erased, mostly in favor of artifacts. Latour strongly stresses that all these "non-humans" are not given, but constructed, all of them are there as results – there is always a process in progress. Even if something like nature is to be found, it will be found eventually, at the end of the process. Getting rid of nature is getting rid of anything considered as given, as being always already there.

Bruno Latour's proposal for facing the environmental crisis can therefore be understood as a way of dealing with technical entities, as dealing with the technical system we have made, but to which we are subordinated as well. His book can thus be read as a solution to the problem of the "autonomy of technology." Technological entities are no longer mere or simple objects: we cannot master them as we would like to. To avoid the subversion of the old dichotomy (the subjects becoming the objects of the objects), we can try to make them our equals, to include them in our social relationships. Thus, Latour's *Politics of Nature* can be seen as a way to answer the "Ellul hypothesis," a solution which, once more, eventually ignores the "White hypothesis," because it relies on a conception of nature that is not only metaphysical, but, first of all, artificial. So putting together, as "non-human," entities formerly considered natural as well as artificial (or technological) is a way to see all of them as artificial.

Not completely so, however. These "non-human" entities retain some features of the natural world: they are strange, disturbing, frightening. If we consider them, it is because they represent a danger. At one point Latour speaks of them as "aliens" (2004, 194) – "strange, frightening creatures from another world"! Nature has not completely disappeared, it has been placed in the beyond, as far away as possible, but it retains its usual characteristic of being exterior, and radically so. What makes non-humans different from humans, and what is common to all of them, is that they partake of this natural feature, exteriority. And this nature is dangerous and frightening, and is different in this aspect from the nature that naturalistic environmental ethics vow to respect. Non-humans are much more frightening than the good nature of the naturalistic environmentalists.

We can draw two conclusions from this outcome of Latour's solution that can help us to answer our question (are naturalistic environmental ethics losing out to technological environmental ethics?):

1. Latour does not really put an end to the modern dualism. He remains dualistic. He merely puts nature much more beyond our reach than the modern constitution used to, and, mainly he does not take the divide between what belongs to nature and what belongs to social communities as constituted once and for all. This divide is a result, an ever-changing result, it has to be negotiated: it is at the end of this procedure that one can know what belongs to communities of humans and non-humans, and what is rejected on the other side, on the side of nature (the outside of the community, where we reject what we do not want to be responsible for) (*Ibid.* 124).
2. If we have problems in dealing with our artifacts, it is because these artifacts are still, in some way, natural, or linked with nature. Therefore it is not possible just to cancel, or delete, nature. Latour's mistake could be to refer only to dualistic relations (nature/culture, nature/artifacts, etc.) and to treat them as if they were identical. They are not: because our artifacts, though man-made, still belong to nature, they are different from other cultural products or forms. Hence we should not rely on a dualistic relation, nor try to get rid of it (one cannot, nature comes back, and fearfully so) but instead we should refer to a three-terms relation, a nature-artifact-culture relation, a triangular relationship in which there is no dominating or privileged vertex.

10.4 Conclusion: Nature Is not a Bygone Reference – We Still Have to Deal with it

1. This implies firstly that an environmental ethics cannot be only an ethics of responsibility. It must include in part an ethics of respect; if we understand respect as being based on the acknowledgement of alterity – alterity meaning not necessarily radical exteriority, but something which stands by itself, which has its own life, a self-sustaining process.
2. Secondly, I agree with Latour's idea that nature is not part of the solution, nature is part of the problem. Nature, in nature conservation, is no longer an unquestionable given. Nature is the object of the debate. For instance, in the current controversy about what to do with wolves, the "naturalness" of the wolf is an important part of the controversy. To say, for example, that "wolves have been reintroduced" is to negate the naturalness of the wolf.

So nature conservation is more and more about what is nature, what is natural, what kind of nature we want to conserve (or preserve). This debate is partly a scientific controversy, between conflicting scientific references: between an ecology of equilibrium and an ecology of disturbances, for instance, between an ecology in which people can only destroy equilibriums, and must therefore remain outside of self-sustaining processes, and an ecology according to which people can enhance biodiversity, so that conserving nature may imply maintaining human activities (such as grazing) in preserved natural areas.

So we can agree with Latour's idea of combining scientific controversies and public political debate, but we would add that these debates and controversies are about nature as well, and that including nature in the debate has ethical implications. It means combining respect with responsibility.

References

Ellul, J. 1954/1964. *The technological society*. Trans. J. Wilkinson, with an introduction by R.K. Merton. New York: Alfred A. Knopf. 1st French ed., published by A. Colin, Paris.

Ferry, L. 1992/1995. *The new ecological order*. Chicago: Chicago University Press. 1st French ed., published by Grasset, Paris.

Gerrie, J.B. 2003. Environmental ethics: Should we preserve the red herring and flounder? *Journal of Agricultural and Environmental Ethics* 16(1): 63–76.

Latour, B. 1991/1993. *We have never been modern*. Trans. C. Porter. Cambridge, MA: Harvard University Press. 1st French ed., published by La Découverte, Paris.

Latour, B. 1999/2004. *Politics of nature: How to bring the sciences into democracy*. Trans. C. Porter. Cambridge, MA: Harvard University Press. 1st French ed., published by La Découverte, Paris.

Serres, M. 1990/1995. *The natural contract*. Trans. E. MacArthur and W. Paulson. Ann Arbor: University of Michigan Press. 1st French ed., published by François Bourin, Paris.

White, L. Jr. 1967. The historical roots of our ecologic crisis. *Science* 155(3767): 1203–1207.

Chapter 11
Epilogue: The Epistemic and Practical Circle in an Evolutionary, Ecologically Sustainable Society

Donato Bergandi

Abstract In a context of human demographic, technological and economic pressure on natural systems, we face some demanding challenges. We must decide 1) whether to "preserve" nature for its own sake or to "conserve" nature because nature is essentially a reservoir of goods that are functional to humanity's well-being; 2) to choose ways of life that respect the biodiversity and evolutionary potential of the planet; and, to allow all this to come to fruition, 3) to clearly define the role of scientific expertise in a democratic society, recognizing the importance of biospheric equilibrium.

In fact, in socio-scientific controversies, which are characterized by complex linkages between some life and environmental sciences objects and economic, political and ethical issues, a posture of transparent, impartial commitment is appearing, more and more, as a deontological necessity.

> The earth is fast becoming an unfit home for its noblest inhabitant, and another era of equal human crime and human improvidence . . . would reduce it to such a condition of impoverished productiveness, of shattered surface, of climatic excess, as to threaten the depravation, barbarism, and perhaps even extinction of the species.
>
> George Perkins Marsh – 1864

> Can human activity really be significant enough to drive the Earth into a new geological epoch? . . . The ultimate drivers of the Anthropocene, . . . if they continue unabated through this century, may well threaten the viability of contemporary civilization and perhaps even the future existence of *Homo sapiens*.
>
> Will Steffen, Jacques Grinevald, Paul Crutzen, and John McNeill – 2011

D. Bergandi (✉)
Department Men-Natures-Societies, Centre Alexandre Koyré,
Muséum National d'Histoire Naturelle, 75005, Paris, France
e-mail: bergandi@mnhn.fr

D. Bergandi (ed.), *The Structural Links between Ecology, Evolution and Ethics: The Virtuous Epistemic Circle*, Boston Studies in the Philosophy of Science 296,
DOI 10.1007/978-94-007-5067-8_11,

> There is a grave danger facing mankind. The danger is not from acid rain, global warming, smog, or the logging of rain forests, as environmentalists would have us believe. The danger to mankind is from environmentalism.
>
> Michael S. Berliner – Ayn Rand Institute 2012

Man has become one of the major factors in evolution on the planet. In this new context, evolutionary, ecological and ethical issues have become some of the more pressing concerns facing humanity. Natural systems are facing steadily increasing human demographic pressure and eco-unfriendly technological and economic activities that risk pushing them beyond their equilibrium and resilience capacities. At the same time, social systems are experiencing a new phase in development. Geographical borders no longer place limits on the communication and globalization of specific cultural and socio-economic patterns. In light of the profound ecological and sociological changes that are rapidly transforming our natural and social environments, humanity has a broad range of choices about what kind of society could be achieved. At the extremes of this spectrum lie various social models that involve very specific man-nature relationships. There is a social model based on the "survival of the fittest," where the struggle for existence between the members of a society, and with the members of other societies, continues to sustain a predatory relationship with the rest of nature. At the other extreme is a more cooperative kind of society, where an enlarged common good is pursued – a common good that is not limited simply to certain short-term specific human interests, that clearly recognizes the right of other non-human species to an existence and that values the persistence of the environments that allow these species to survive and proliferate.

Some of the challenges that our societies must face in the coming years are more pressing than others. First, in the context of the continuous decline in the planet's biodiversity, we must decide whether to structure our societies around management policies and patterns that ensure the preservation of nature, or its conservation. Second, among the various options for social and economic development, we need to choose a model that is consistent with respect for the greatest evolutionary potential of our planet. Third, we must clearly define the role of scientific expertise in a democratic society that recognizes as one of its objectives the maintenance of a biospheric equilibrium favorable to biodiversity.

The choice of the environmental trajectory of our societies is no trivial matter: it will have direct consequences on our ecosystem management strategies. The traditional opposition between preservationism and conservationism, emblematically represented by the tensions and even clashes between the ideals of John Muir and those of Gifford Pinchot (Bergandi and Blandin 2012), implies very different types of societies, and consequently very different man-nature relationships. The epistemic and practical center of gravity of preservationism is nature and its equilibrium, and more particularly, the wilderness, a state of nature where evolutionary processes can come into existence without encountering any major hindrances from human activity. A society structured around a preservationist worldview will limit the impact of man on natural systems to a minimum. Such a society will recognize the ontological integration of humanity in nature and will ensure that technological and economic development is congruent with nature's dynamics. In other words, from the beginning, technology and the economy would be thought of as constituent parts of the environment, meaning that one primary objective of the way they are developed would be their coherent adaptation within natural systems.

The barycentre of conservationism is human society and its economic order, which, in the best scenarios, will make "wise use" of natural resources. From this perspective, nature is synonymous with natural resources: nature is nothing but a reservoir of goods and potentialities for developing our economies. Man is at the center of the ecosystems, and everything is thought and lived in function of human interests and expectations. Biodiversity is conserved because its decline can be harmful to the well-being of humanity and not because destroying other species, and environments, and disrupting biospheric equilibriums to increase human consumption is ethically reprehensible in itself. Such a conservationist perspective on the man-nature relationship underlies the sustainable development paradigm, which tends to be assumed as a guiding principle by governments, international institutions, corporate business, civil stakeholders and the public. Its widespread acceptance undoubtedly results in part from the vagueness and flexibility of its semantic core, which allows different stakeholders to see in this developmental model whatever they want to see in terms of their specific interests and purposes. In reality, the sustainable development paradigm has not yet won the competition for the governance of the planet. It is only one of the models that holds the stage, sharing political space with the economic de-growth movement and various models of environmental ethics.

Formally, sustainable development is a model with a high moral content. From a reading of international treaties, conventions and declarations, the prospective planetary society, grounded on the globalization of sustainable development policies to the totality of human societies, purports to be a truly democratic society where everyone, regardless of their gender, age or social condition, can satisfy their basic needs. It is a social model where awareness of the ontological interdependence between man and nature determines a minimal impact of human activity on ecological systems, entailing profound lifestyle changes and a substantial reorientation of consumption patterns. In such a society, an awareness of the priority of the common good would guide the actions of individuals and economic groups (WCED 1987).

In reality, the sustainable development model is proposing a future world that would be balanced precariously between utopia and Janus-like postures. In fact, its peculiarity consists in sustaining antinomic positions: it simultaneously supports the preservation *and* the conservation of nature; it recognizes the instrumental *and* intrinsic value of biodiversity (see Preamble of the Convention on Biological Diversity, CBD and UNEP 2001); and it sustains continued economic growth in a finite world of finite natural resources. Even if this growth is called "sustainable", this takes for granted the possibility of achieving the harmonious coexistence of ecological sustainability and economical sustainability. Moreover, it does all that without fundamentally calling into question the mainstream productivist economic model of development. The utopian side of the model consists in imagining a new international political and economical order grounded on the planetary globalization of democratic governance, which is to be better adapted to guarantee the common good of the populations against private, particular interests and more inclined to respect environmental integrity. Citizen participation in environmental decision-making would represent a barrier against unsustainable uses of the environment. Nevertheless, the ambiguity intrinsic to a utopia, as a good place or as a totally non-existent place, is also intrinsic to the sustainable development model, and this represents, at the same time, its strength and its weakness.

The magnificent, soothed, idealized world represented in international treaties and conventions is in stark contrast to the actual results of implementing sustainable development programs and policies. If the sustainable development model's internal contradictions and the gap between the ideal and the reality are not overcome, then it will long remain in the empyrean of the utopias. It could perhaps endure as a kind of inefficacious religious mantra, or a very efficacious instrument of intoxicating advertisements to promote the idea that everything has changed, whereas, in reality, the development taking place is incompatible with the imaginary world of "sustainability." Ultimately, the sustainable development model's internal contradictions will likely determine its fiasco.

Among the key challenges that we have to face in the current context, clarification is needed about scientists' role in society, their margins of autonomy or their dependence on the rest of society, and the role of internal factors (logical, methodological) and external factors (social) in the determination of controversies in the life and environmental sciences.

In fact, nowadays, as at the beginnings of modern science, scientific knowledge is rarely free of social interests or of practical and cultural repercussions on the life of society. Some controversies sprang out of the meeting between scientific questions and moral, economic and religious worldviews and ways of life. Darwin's scientific proposal about the causes of the variability of species is an historical and emblematic example of the comings and goings between science and society. With Darwin, from the beginning a "no religion's land" has been instantiated. As grounding for the scientific neutrality of his position towards religion, Darwin cautiously extracts one specific aspect of Whewell's natural theology concerning the role of the Laws of Nature. Whewell considered that the universe was the work of an omnipresent Deity, and that the laws of nature were the expression of its power. But Darwin chose as a frontispiece quote for *The Origin* exactly the passage where Whewell clearly indicates that nature is ruled by laws and not directly by specific creation events of the Divine power.

> But with regard to the material world, we can at least go so far as this—we can perceive that events are brought about not by insulated interpositions of Divine power, exerted in each particular case, but by the establishment of general laws (Darwin 1859; Whewell 1833, 356).

The existence of a God, legislating or not legislating for nature, is not the object of Darwin's research; such a question is explicitly dismissed as a metaphysical, "insoluble" non-scientific issue (1876, 73): the study of nature and the study of religious beliefs belong to different realms of meanings. The solution chosen by Darwin, *i.e.* neutral impartiality about a question that is considered metaphysical, and not scientific, is not always an easy path to follow. Our societies are living through major cultural transformations, and new challenges are increasingly emerging from the relationship between science and society. More particularly, some of these challenges concern the life and environmental sciences and the public understanding of scientific knowledge and its applications.

Some controversial scientific topics lying on the border between science and society (*e.g.* the conservation of biodiversity; the ecological and social impact of certain new bio-technologies; and climate and environmental change issues, among others), involve research dynamics that are inevitably intermingled with the personal philosophical, ethical and political, conscious or unconscious, convictions of the researcher. The answers to socio-scientific questions like these cannot be decided exclusively on the basis of so-called impartial scientific results. These questions belong to an order of meaning that is totally different from questions such as: does the universe have borders? Do neutrinos move faster than the speed of light? In the case of the socio-scientific controversial issues, it is the whole universe of values, of the researcher's moral and political convictions, that is involved, because these issues are not exclusively scientific but also economic, political, ethical and cultural.

With socio-scientific issues, with topics that clearly have economic, ethical and political implications, what is at stake is the epistemic and sociological autonomy of the scientific community. The scientific ideal of a clear dichotomy between scientific facts and values, between science and ethics – historically grounded on positivistic and neo-positivistic perspectives – is still pervasive today, with different degrees of intensity, depending on the scientific community of reference, and this dichotomy still grounds our current scientific rationality. Science, with its descriptive statements, referring to the "facts," produces meaningful sentences which, being empirically testable, are the expression of objective knowledge. Ethics, with its prescriptive statements – non-testable, relative and subjective – does not accord with the ideal of objective knowledge. In fact, from this perspective, the possibility that ethical sentences could be true or false is denied: because they are not factual statements, they are in that regard nonsensical, meaningless. The assumption of a dichotomy between fact and values is no longer considered crystal-clear, or evident at all (Putnam 2002).

The treatment of this dichotomy substantially parallels the treatment of another classical dichotomy, *i.e.* between theory and observation. The experimental and observational dimensions of science do not exist in a conceptual vacuum. The ideational, theoretical dimension, even in the pre-theoretical form of hypothesis, directly participates in the definition of scientific problems, "legitimate facts" and solutions. Ideas permeate scientific facts; they allow facts to emerge as such. Facts, as Dewey reminds us (1986, 127 [1938]), are not "given" (by our senses or observation methods) but, rather, are "taken" (extracted from the complex, total field of the problematic situation by the ideational contents of the scientific inquiry).

Using a comparable constructivist perspective, a functional, reciprocal correlation between facts and values can be found. Dewey tells us that (epistemic and ethical) values are not fixed, isolated ends-in-themselves, but they are ends-in-view, ends to be attained, reached, plans of action and purposes. They directly participate in the discrimination, or selection, of the means used to carry out scientific inquiry and the selection of data. Values determine the elements to be taken into consideration in the formation and the framing of a problem, and they should be understood, and lived, as hypotheses that must be practically evaluated and socially tested. Finally, more specifically, values can play a role in encouraging and orienting scientific research; they can become guides for observation or guiding principles for scientific work (Dewey 1986, 491 [1938]; see also: Popper 2004, 16 [1935]; Einstein 1993, 28 [1934]).

The reality of research is grounded in an entanglement of facts and values, independently of whether controversial scientific issues are involved. Nevertheless, the original, ontological and epistemological mix of facts and values assumes all its significance in the case of socio-scientific controversies. In fact, when the researchers are confronted with topics about which they are, consciously or unconsciously, committed – from an affective, ethical or political point of view – how can they not reject, dismiss or undervalue, hypotheses, theories, or quite simply "facts" that are not congruent with their own worldviews? Once the idea is accepted that values permeate facts – even if the search for impartial, objective knowledge continues to represent the ultimate aim of any scientific community – it follows that, however much individual researchers strain to achieve complete objectivity, they never succeed. Nevertheless, the entirety of the results of the scientific community, over time, will increasingly approach objective knowledge, without ever totally reaching it at a given time.

At best, the scientist aware of the multiple forms of his commitment relative to the object of research will try to set aside prejudice and critically evaluate the data, theories and values at stake. To avoid any eventual misdirection in the research and in the communication of the results of the research with lay people, one possible posture to be applied could be that of an impartial commitment. In that situation, in the case of the socio-scientific issues, the researcher who keeps in mind, as much as possible, an impassible, ataraxic posture relative to the theories and values situated at the antipodes of his worldview will explicitly set out and clarify his ethical and political preferences. Considering all the complex linkages between some life and environmental science objects and economic, political and ethical issues, a posture based on this kind of transparency is appearing, more and more, to be a deontological necessity. Otherwise, non-neutral scientific results will be presented as the expression of an objective knowledge, masking their true appearance. The ensuing confusion would have decisive repercussions on our political conduct and our choices about what type of planet we want for us, now, and for future generations.

An impartial, enlarged scientific community, involving not only natural scientists, but also philosophers, historians, anthropologists, sociologists, and economists, among others, should commit itself to developing a collective awareness of the consequences of human activity on the planet. A feedback process that can reverse the process of environmental degradation could be activated if, and only if, it is the expression of a refoundation of the value-systems that underpin cultures and societies that today are forging the "biosphere" in an "environment" that reduces non-human nature to a simple extension of the human species. In the future, there is a real possibility that evolutionary biology and ecology will develop their research by no longer using natural systems as the object of study, but agro-ecosystems managed to optimize the production of products functional to the subsistence of the human species. In this case, the Biosphere will be very different from now; it would be closer to an Urbosphere, an artificialized, urbanized planet embellished by some patches of nature here and there. The ethical and political-economic preferences that we are making today will determine the evolutionary and ecological paths of the future of the planet. These choices will directly produce the environments where natural selection processes will operate, selecting, if we make bad decisions, more and more domesticated species.

Our systems of values are selected by the environment through the consequences of our activities on natural systems. Our economic activities are the embodiment of our ethical values, and the repercussions that these activities have on our lives, on other species and on the environment represent, in a figurative way, the biosphere's refusal or approval of the values underlying these activities. Among the many ethical options which humanity has at the moment is a farsighted co-evolutionary ecological ethics. This option, which pays equal attention to the prosperous evolutionary flourishing of both natural and social systems, is based on a fundamental worldview reversal. To paraphrase the well-known anthropocentric sentence of Baxter (1974; see in this vol., 1st Chap., pp. 6–7), we could say that, what is good for nature, penguins and pine trees is, in many respects, good for humans. Natural equilibriums are the basis of life on this planet, and, even embracing a more restricted, blind, short-term homocentric point of view, their possible breakdown will in no way be functional to the interests of the human species. "Man is Nature becoming conscious of itself," held Elisée Reclus (1905, I; see also Bergandi 1998, 525–529), one of the forerunners of human geography who clearly recognized that the study of nature is the precursor of action to preserve animal and plant species. A humanity like this, a conscious expression of nature, must definitely decide on its place in the world. We may continue to consider ourselves as the acme of evolution, the master and possessor of nature or, finally, we may recognize ourselves to be a co-evolutionary entity that is strictly integrated with the rest of nature.

If this recognition is accompanied by a cultural, ethical transition involving the endorsement of the sharing, at least in principle, of intrinsic value – traditionally considered as a unique, typical human property – with our other travel companions in the biosphere, the other actors (other species) and scenes (environments) in this evolutionary journey, then humanity will, presumably, increase its chances of saving itself. As humans, to survive and develop our potentialities, we have no solution other than to metabolize environmental energies, but we can, and likely must, minimize our impact on the biosphere's evolutionary and ecological processes. By doing so, by respecting non-human nature for its own sake, grounded on the very fact of its existence as an evolutionary entity, we will, most likely, witness a paradoxical side effect: that we too will continue to form a part of this biosphere for a long time to come.

References

Baxter, W.F. 1974. *People or penguins. The case for optimal pollution*. New York: Columbia University Press.

Bergandi, D. 1998. Geography of human societies. In *The European origins of scientific ecology*, ed. P. Acot, 521–533, 2 vols. + cd-rom. Amsterdam: Overseas Publishers Association/Gordon and Breach Publishers/Editions des Archives Contemporaines.

Bergandi, D., and P. Blandin. 2012. De la protection de la nature au développement durable: genèse d'un oxymore éthique et politique. *Revue d'histoire des sciences* 65(1): 103–142.

CBD and UNEP. 2001. *Handbook of the convention on biological diversity*. London: Earthscan Publications.

Darwin, C. 1859. *On the origin of species*. London: John Murray.

Darwin, C. 1876. *Recollections of the development of my mind & character*. Autobiography [1876-4.1882], 1–121. Transcribed by Kees Rookmaaker. http://darwin-online.org.uk/

Dewey, J. 1938/1986. *Logic: The theory of inquiry*. New York: Holt, Rinehart and Winston. Repr., *The later works of John Dewey 1925–1953*, vol. 12, 1938, ed. J.A. Boydston, ix–xxvii, 3–793. Carbondale and Edwardsville: Southern Illinois University Press.

Einstein, A. 1993. *The world as I see it*. Trans. Alan Harris. New York: Citadel Press. Orig. ed. *Mein Weltbild*. 1934.

Marsh, G.P. 1864. *Man and nature; or, physical geography as modified by human action*. New York: C. Scribner.

Popper, K.R. 1935/2004. *The logic of scientific discovery*. London/New York: Routledge.

Putnam, H. 2002. *The collapse of the fact/value dichotomy and other essays*. Cambridge MA/ London: Harvard University Press.

Reclus, E. 1905. *L'Homme et la terre*, vols. I–VI. Paris: Librairie universelle.

Steffen, W., J. Grinevald, P. Crutzen, and J. McNeill. 2011. The anthropocene: Conceptual and historical perspectives. *Philosophical Transactions of the Royal Society A* 369: 842–867.

WCED. 1987. *Our common future*. Oxford: Oxford University Press.

Whewell, W. 1833. *Astronomy and general physics considered with reference to natural theology*. London: William Pickering.

Author Index

D. Bergandi (ed.), *The Structural Links between Ecology, Evolution and Ethics: The Virtuous Epistemic Circle*, Boston Studies in the Philosophy of Science 296,
DOI 10.1007/978-94-007-5067-8,

Subject Index

D. Bergandi (ed.), *The Structural Links between Ecology, Evolution and Ethics: The Virtuous Epistemic Circle*, Boston Studies in the Philosophy of Science 296,
DOI 10.1007/978-94-007-5067-8,

H

I

U

V

W

Y

Z

Printed by Publishers' Graphics LLC